# MathFINDER™ Sourcebook:

## A Collection of Resources for Mathematics Reform

To use independently or in conjunction with the
MathFINDER CD-ROM

written and edited by
Laurie Kreindler and Barbara Zahm

The work upon which this publication is based was performed pursuant to
Grant No. MDR-8955079 from the National Science Foundation to Education
Development Center, Inc. Opinions expressed are those of the authors and
not necessarily those of the foundation.

Copies of this book are available from:

**The Learning Team**™
Box 217
Armonk, New York 10504-0217
Tel   914/273-2226
Fax   914/273-2227

# Table of Contents

# Acknowledgments

We at The Learning Team would like to thank the many people who have been involved in the creation of the **MathFINDER Sourcebook.**

To begin with, we are grateful to Tom Berger and Joe Adney of the National Science Foundation. Their encouragement, enthusiasm, and support have been very much appreciated.

Next, a very special acknowledgment must go to all those at the Education Development Center, Inc. (EDC), whose help on this book has been invaluable. As the creators of the **MathFINDER CD-ROM: A Collection of Resources for Mathematics Reform** (developed from the Resources for Mathematics Reform Project), EDC has been especially important in all aspects of the production of the **MathFINDER Sourcebook.** The important link between the **MathFINDER CD-ROM** and the **MathFINDER Sourcebook** has meant that we have greatly benefited from the labor that went into the creation of the **CD-ROM.** EDC provided day-to-day guidance and oversight which has been indispensable. We are especially grateful to Mark Driscoll, whose extensive background in mathematics curricula, understanding of the *NCTM Standards,* and commitment to excellence has made him an outstanding partner on this project. Sherry Booth, Sheila Flood, Grace Kelemanik, D. Midian Kurland, and Leigh Peake have each contributed greatly. We would especially like to thank Deborah Bryant for her gentle patience with all of our questions as well as her conscientious review of each page of the **MathFINDER Sourcebook.**

We are also grateful to the members of the Resources in Mathematics Reform Oversight Committee (listed on the inside back cover) for their good advice, careful review, and especially for their contribution to the selection process of the curricula.

In addition, we would like to thank all of the authors and publishers included in the **MathFINDER Sourcebook** for their cooperation and input.

Also providing valuable feedback and mathematics consultation were Ariane Cherbuliez, Susan Isenberg, and Deborah Stampfer. Jennie Allen carefully copy edited each page. We would also like to thank Susan Carucio, Kim Kourtakis, Kirtine Lee, Betsy Pennebaker, and Janae Williams for their skillful writing and editorial assistance.

John Nordland provided crucial support, creative ideas, and design expertise throughout. He worked with us on the logo and design of the cover as well as the layout of each page. Larry Atwood, from Arcata Press, and Kate Molitar have been both informative and supportive.

Finally, we would like to thank Barbara Zahm, our co-editor and writer, for bringing to this project her talent, diligence, and dedication.

**The Learning Team**
Tom Laster - Executive Director
Laurie Kreindler - Creative Director

# Letter to the Educator

Dear Educator,

Welcome to **MathFINDER Sourcebook: A Collection of Resources for Mathematics Reform.** This book will help you join the exciting reform movement in mathematics education which has led to the creation of the *Curriculum and Evaluation Standards for School Mathematics* (the *Standards)* by the National Council of Teachers of Mathematics. The *Standards* express the national professional consensus for the direction of mathematics teaching in the next decade. Our intent was to create a handbook that could direct teachers towards existing curricula that illustrate the educational goals established by the *Standards.*

The **MathFINDER** project was developed in response to the need of mathematics educators to locate curricula with which to implement the principals and goals outlined in the *Standards.* After thoughtful investigation, leaders in the field felt that many of those materials already existed. During the nineteen-sixties and nineteen-seventies many curriculum programs had been developed. These dispersed curriculum programs were collected and organized within the *Standards* framework and are presented here to help the mathematics reform movement come alive.

- The **MathFINDER Sourcebook** is a thorough introduction to the *Standards.* The text and organization of the *Standards* have been incorporated into the **MathFINDER Sourcebook** as its basic structure.

- For those already familiar with the *Standards,* the **MathFINDER Sourcebook** provides parts of lessons which illustrate each of the 40 Standards: 13 Standards for K–4, 13 Standards for 5-8, and 14 Standards for 9-12.

- The **MathFINDER Sourcebook** also introduces 30 curricula, selected by a panel of experts who are national leaders in mathematics education, to illustrate the recommendations of the *Standards.*

We at **The Learning Team** are also pleased to be able to offer the **MathFINDER CD-ROM: A Collection of Resources for Mathematics Reform,** developed at the Education Development Center, Inc. This remarkable new disc provides computer access to approximately 1100 classroom lessons selected for their outstanding quality. It is a virtual library at your finger tips. The **MathFINDER Sourcebook** and the **CD-ROM** can be used independently or in conjunction with one other.

We encourage everyone to read and study the entire *Standards* document. This **MathFINDER Sourcebook** is not a substitute for the *Standards* but a complement to them.

The **MathFINDER Sourcebook** and the **CD-ROM** are important new tools for mathematics reform. We hope they will spark your ideas, save you time, and lead to more effective teaching. Please give us your feedback. We value your ideas.

Sincerely,

**The Learning Team**

# Introduction

The **MathFINDER Sourcebook** is divided into two sections.

## Section I - The *Standards* and Examples for Their Application

Following the structure of the *Standards,* this section is broken into the three grade-level groups: K-4, 5-8, and 9-12. Each Standard is reproduced in full, with its subparagraphs marked with bullets. A short synopsis of the focus and discussion found in the text of the *Standards* follows. For each Standard, two illustrative lessons are presented. In some cases the lessons are appropriate to the entire Standard and in other cases to one specific aspect, or bullet, of a Standard.

## Section II - The Curricula

Section II gives a brief description, a short history, and information for ordering each of the 30 curriculum programs from which the lessons in Section I were selected. Each curriculum program is cross-referenced for easy access to the lessons.

## Ways to use this book:

### As an introduction to the *Standards*
The **MathFINDER Sourcebook** complements the *Standards* and can help link the *Standards* to classroom use. The **MathFINDER Sourcebook** provides excerpts from specific classroom lessons to demonstrate the approach to mathematics suggested by each Standard. In some cases you may find that the lessons we have selected are right for you but in other cases you may know of lessons that are more applicable.

### As an in-service and pre-service tool
The **MathFINDER Sourcebook** is an aid to any teacher-training program designed to promote mathematics reform and the implementation of the *Standards*. Discussions of how selected lessons illustrate a particular Standard are a natural starting point for teachers trying to incorporate the *Standards* in their classrooms. Among the questions that might be discussed are: How do these lessons illustrate the Standard? Which lessons in the **Sourcebook** other than the ones selected could be used to illustrate different Standards? How would curricula other than those presented here implement the goals and approach established by the *Standards*?

### As an aid to curriculum selection and library acquisition
The **MathFINDER Sourcebook** will serve as a reference book for math specialists, department heads and librarians interested in purchasing curricula that have a *Standards* approach, as well as for classroom teachers most directly concerned with instruction.

# Section I The *Standards* and Examples for Their Application

**Grade level K—4**

*Problem solving is not a distinct topic, but a process that should permeate the mathematics curriculum. Both of the lessons below encourage students to use a variety of problem-solving techniques. Students should develop and apply strategies as well as verify and interpret their results.*

## Teaching Integrated Math and Science (TIMS)

### The Bouncing Ball

"The Bouncing Ball" illustrates how a problem-solving approach can create a rich learning environment in the classroom.

Working in teams of two, students learn about ratio, proportion, and quantitative variables by dropping tennis balls and super balls, and then measuring their respective bounce heights. Analyzing the data, students discover a linear relationship between the variables.

Using graphs that they create, students investigate problems such as: If the drop height of the tennis ball is 60 cm, what is the bounce height? If the drop height of the tennis ball is 160 cm, what is the bounce height? If the tennis ball rebounds to a height of 55 cm, from what height was it dropped? Other prediction questions involving proportional reasoning can also be asked.

TIMS materials encourage a problem-solving classroom approach with an experimental emphasis. The materials for each of its 80 experiments include: Teacher Lab Instructions; Teacher Lab Discussion pages; Experiment pages; Data Tables and TIMS Tutors which provide information on the scientific concepts.

- For further information about TIMS turn to pages 22, 47, 54 and 114.

From TIMS, The Bouncing Ball, Experiment Unit 201. TIMS materials were created at, and are available from, the project office at the University of Illinois at Chicago.

---

# The Bouncing Ball

## Teacher Lab Discussion

### Overview

This is one of the first experiments involving two quantitative variables. Both are length variables, the simplest we can use. In this experiment we will determine the exact relationship between the height from which a ball is dropped and the distance that it rebounds.

### Choosing the variables

One of the first tasks in designing an experiment is deciding what the two variables, manipulated and reponding, will be. A ball is one of the simplest and most familiar objects in the child's world. One can think of many variables associated with a ball. In this experiment we have selected two interesting

Fig. 1

quantitative variables. **Fig. 1** depicts a student's picture of the **Bouncing Ball** experiment. The two variables are the release height (measured from the bottom of the ball, in cm) and the bounce height (again measured from the bottom of the ball in cm). Suzy releases the ball and her partner, Lynn, kneeling to get a level view of the ball, measures the bounce height. Notice that both variables have been labeled on the picture. This is an essential part of the experiment. The next step in the experiment is constructing a data table so we can enter the ex-

perimental data in an organized and logical way. In **Fig. 2** we show you the blank data table. First we want to see how to choose the values of the manipu-

| Type of Ball _____ | | | | |
|---|---|---|---|---|
| D (cm) | B (cm) | | | |
| | Trial 1 | Trial 2 | Trial 3 | Average |
| 40 | | | | |
| 80 | | | | |
| 120 | | | | |

Fig. 2

lated variable. Later we will discuss the method of obtaining accurate values of the responding variable.

### Choosing the values of the manipulated variable

In this experiment, the release height is the manipulated variable. You, the teacher, can pick any values that you want. However, a little common sense shows us that some choices are better than others. One goal of any experiment is to see a pattern in the relationship between the variables. We must select at least three values of the manipulated variable in order to see the relationship between the variables. Also, if the values you choose are too close together you may not be able to see the pattern. For example, dropping the ball from 20 and 22cm, there may be no significant difference in B because of the "error" in making the measurement. It is not easy to determine the maximum height of a moving bouncing ball with better than 2 or 3cm accuracy. So then, what values of D should we pick? It is best to pick them so that they are multiples of

The *Standards* state that in grades K–4 the study of mathematics should emphasize problem solving so that students can–

- Use problem-solving approaches to investigate and understand mathematical content;

- Formulate problems from everyday and mathematical situations;

- Develop and apply strategies to solve a wide variety of problems;

- Verify and interpret results with respect to the original problems;

- Acquire confidence in using mathematics meaningfully.

# Unified Science and Mathematics for Elementary Schools (USMES)

## 1. LOG ON SOFT DRINK DESIGN

by Mary Lou Rossano*
Hardy School, Grade 2
Arlington, Massachusetts
(October 1973–June 1974)

ABSTRACT

*This second-grade class spent approximately two hours per week from October to June working on a Soft Drink Design challenge to develop a snack-time drink that everyone in the class would like. Unable to duplicate drinks they had made during random mixing, the children realized the importance of accurately measuring ingredients and recording recipes. They devised a class questionnaire on drink preferences, tallied the results on a master chart, and then worked for several months in four groups to develop drinks that fit the survey criteria for a popular drink. Student interest in Soft Drink Design was so great that, on their own, the children began drawing pictures and writing stories about their activities, and several even stayed after school to work on a mural. When one drink was voted the most popular, the students decided to sell it to other classes. The Mixing Group scrounged and cleaned quart jars for mixing and bottling large quantities of "Super Grape," and they devised an order form. The Advertising Group developed slogans, made posters and announcements, and designed labels for bottles. The Drink Stand Group had difficulty obtaining supplies for their stand and ended up constructing a table for the classroom. Before selling their drink, the class used pie graphs to perform a cost analysis and figure profit per bottle. Sales of Super Grape continued for six weeks. The second graders used half their profits to buy a plant for a nearby nursing home and half to treat everyone in class to ice cream cones.*

## Soft Drink Design

"Soft Drink Design," like all of the USMES lessons, provides an in-depth problem-solving experience. In this unit, students are challenged to invent a new soft drink that would be popular and could be produced at a low cost.

The first problem to be solved is where to begin. After group discussion, some students may decide to randomly mix drinks and record favorite recipes. Others might conduct surveys to determine preferences for flavor, color, temperature, and degree of carbonation. Results are tallied and displayed on bar graphs and histograms. Once the class has selected a recipe, students use comparative shopping and cost analysis to find the best buys for ingredients. Finally, in groups, students determine production cost, selling price, profit, and expected volume of sales.

Teacher support materials include logs from teachers who have previously taught the unit (as shown here).

- For further information about USMES turn to pages 31, 55 and 115.

From USMES, Soft Drink Design. USMES materials were created at Education Development Center. For further information contact The Learning Team.

*Mathematics can be thought of as a language which plays an important role in helping students link their informal, intuitive ideas with the abstract symbolism of mathematics. These two lessons encourage students to listen to, describe and write about mathematical ideas.*

## Journeys in Mathematics

### A Letter to the King

"A Letter to the King," like many of the Journeys activities, provides a creative setting for students to read, write and draw about mathematics.

This activity focuses on developing students' measurement and estimation skills. Students read a chapter in *Gulliver's Trunk*, a story based on the first part of Jonathan Swift's *Gulliver's Travels*. They read first for content, and then for clues as to the size of the objects in the lands Gulliver visits: for instance, an adult human in Lilliput is 6 inches tall and a horse in Lilliput is 5 inches tall.

After recording their clues in a journal, students measure objects in the real world and write a letter to the king of Lilliput making comparisons such as a Lilliputian adult is about as tall as the length of a pen in Ourland. Students write stories and use their measurement and estimation skills to draw pictures about their own imaginary visits to Lilliput.

- For further information about Journeys in Mathematics turn to pages 24, 49 and 98.

From Journeys in Mathematics, <u>My Travels with Gulliver</u>, Activity 2: "A Letter to the King," page 33. Journeys in Mathematics materials were developed by Education Development Center and are available from WINGS for learning.

---

### Activity 2
### A Letter to the King

**Write a Letter to the King of Lilliput**

*We would use inches to measure most things in Lilliput, since that is what we use to measure small stuff in Ourland. We also use feet and yards when things are bigger. There are 12 inches in a foot and 3 feet or 36 inches in a yard.*

*I'm 4 feet 6 inches tall. That is small here, but I bet that is as tall as your castle. No one would call me 'Shorty' in Lilliput!*

Students read Chapter 2 of *Gulliver's Trunk*. Tam, Shaun, and Aunt Linnea find a letter about Gulliver's adventures in Lilliput. This letter is based on the first part of Jonathan Swifts' *Gulliver's Travels*. By looking for clues in the letter, students make many discoveries about the sizes of people and things in Lilliput.

The king of Lilliput wants to know how big things are in Ourland and how we measure them. Students measure objects in Ourland and look for objects that are about the same size as objects in Lilliput. Then they write a letter to the king of Lilliput explaining the English system of measurement. They also write stories or draw pictures about their own imaginary visits to Lilliput.

The *Standards* state that in grade K–4 the study of mathematics should include numerous opportunities for communication so that students can–

- Relate physical materials, pictures and diagrams to mathematical ideas;

- Reflect on and clarify their thinking about mathematical ideas and situations;

- Relate their everyday language to mathematical language and symbols;

- Realize that representing, discussing, reading, writing, and listening to mathematics are a vital part of learning and using mathematics.

# Madison Project

## Postman Stories

"Postman Stories" is an example of a Madison Project lesson which uses children's real world experiences to develop meaning for abstract mathematical concepts.

In this lesson, students discover the rules of the arithmetic of signed numbers by reading a story about a postman who brings and takes away checks and bills.

In the transactions, receiving mail is represented by addition and having mail taken away is represented by subtraction. The value of a check is represented by a positive number and the value of a bill is represented by a negative number. Students create mathematical sentences based on the receiving or sending of mail (as shown in the example). They make up postman stories corresponding to a problem, and create problems that correspond to a story.

- For further information about the Madison Project turn to pages 14, 46, 73, 85 and 101.

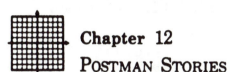

Chapter 12
POSTMAN STORIES

[page 21]

We would like to invent an arithmetic for numbers with signs.

To do this, we look at an example.

Suppose a postman brings you a check for $3. We can represent this as $^+3$. If he brings you a bill for $2, we can represent that as $^-2$.

(1) Suppose the postman brings you a check for $5 and a bill for $3. Are you richer or poorer? By how much?

Can you make up a postman story for each problem? What answer do you get for each problem?

(2)     $^+2 + {}^+4 = ?$

From the Madison Project, Discovery in Mathematics, Chapter 12: "Postman Stories," pages 95-98. Discovery in Mathematics was written by the Madison Project and is available from Cuisenaire Company of America.

*The development of logical thinking and reasoning skills is fundamental to mathematics learning. A classroom climate should be established that places critical thinking at the heart of instruction. The lessons below encourage students to explain, justify, and elaborate on their thinking.*

# Lane County Mathematics Project

## Number Riddles

Lane County materials put a special emphasis on developing students' problem- solving skills. "Number Riddles" provides a series of exercises which help students develop critical thinking skills, including conjecturing and validating.

Students are given a set of clues about a mystery number and are asked to determine the number described. In solving these riddles, students develop several problem-solving skills. They learn to satisfy one condition at a time, recognize restrictions, eliminate possibilities, and discard extraneous information. In the latter part of the activity, students create riddles of their own by making up sets of clues for numbers that satisfy certain conditions.

- For further information about the Lane County Mathematics Project turn to pages 17, 66 and 99.

From the Lane County Mathematics Project, Problem Solving in Mathematics, Grade 4, Section 9, "Number Riddles," pages 385-86. The Problem Solving in Mathematics series was created by the Lane Education Service District and is available from Dale Seymour Publications.

---

### NUMBER RIDDLES

1. Judy Jay made up some number riddles. Solve each.

   a. I am thinking of a number.
      - It is odd.
      - It is between 1 and 100.
      - It is higher than 20.
      - It is smaller than the answer to 6 times 6.
      - It is a multiple of 5.
      - The sum of its digits is 7.

      What is the number? _____

   b. I am thinking of a new number.
      - It is a multiple of 3.
      - It is not even.
      - It is greater than 20.
      - It is lower than the answer to 7 x 6.
      - The sum of its digits is even.
      - The two digits in the number are the same.

      What is the number? _____

   c. I am thinking of another number.
      - It is greater than the answer to 5 x 10.
      - It is smaller than 100.
      - It is even.
      - It is not 70 or less.
      - It is not a multiple of 4.
      - It is not a multiple of 3.
      - It is less than 80.

      What is the number? _____

2. Make up your own set of clues for a number less than 100.

3. Make up a set of clues for a number between 1 and 200. Use at least 6 clues.

**The *Standards* state that in grades K–4 the study of mathematics should emphasize reasoning so that students can–**

• Draw logical conclusions about mathematics;

• Use models, known facts, properties, and relationships to explain their thinking;

• Justify their answers and solutions processes;

• Use patterns and relationships to analyze mathematical situations;

• Believe that mathematics makes sense.

# Used Numbers:
# Real Data in the Classroom

## *RAISINS AND MORE RAISINS*

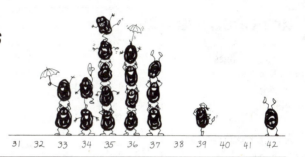

31  32  33  34  35  36  37  38  39  40  41  42

### INVESTIGATION OVERVIEW

#### What happens

Students count the number of raisins in small boxes of raisins (one box for each student). They share methods of organizing, examining, and interpreting their data. Following this activity, students compare other sets of data (provided) from different samples of small boxes of raisins. They invent ways to compare data and also use the median as a way of making comparisons. Students use line plots as well as their own representations to help them describe and compare distributions of raisins.

The activities take two class sessions of about 45 minutes each.

#### What to plan ahead of time

▼ Anticipate students' inclination to use the averaging algorithm, but discourage

them from using it. For suggestions on helping students focus on describing the data, rather than using the algorithm, read the Teacher Note, *The need for data description* (page 14).

▼ Provide small boxes of raisins (the half-ounce size), at least one for each student (Session 1). Have five extra boxes available for Session 2.

▼ Provide small stick-on notes and unlined paper for making sketch graphs (Session 1).

▼ Become familiar with making a line plot and finding a median. See the Teacher Notes, *Line plot: A quick way to show the shape of the data* (page 17) and *Finding and using the median* (page 18).

▼ Duplicate Student Sheet 1 (page 79) for each small group (Session 2).

#### Important mathematical ideas

**Describing the shape of the data.** While the focus of this unit is on finding and using "middles" in the data, describing the shape of the data is a critical and integral part of this work. Later, the *mean, median,* or *mode* will provide an important "summary statistic," but students must first understand what is being summarized. Students need to describe and find important features in the data. Help them gradually move from noticing individual features of the data ("Two boxes had 33 raisins, three boxes had 34 raisins") to describing the overall shape of the distribution ("Over half of the boxes had between 34 and 37 raisins").

**Making quick sketches of the data.** Graphs and tables are used not just for a final presentation of results, but also as working tools to represent data during the process of

Raisins and more raisins    13

## Raisins and More Raisins

"Raisins and More Raisins" illustrates how a mathematics exercise can foster a spirit of inquiry in the classroom. Through this lesson, students begin to develop their critical thinking as well as their estimation skills.

In this lesson, each student is given a small box of raisins and asked to estimate the

number of raisins in the box. After several students have offered their ideas, the students open their boxes and count just the top layer of raisins. Using this new information, students work in small groups, revising their estimates, asking each other questions, and giving reasons for their ideas. Finally, they count the number of raisins in each box, and record and graph the results. This step-by-step process helps students develop their analytical skills.

• For further information about Used Numbers turn to pages 33 and 119.

From Used Numbers, Statistics:Middle, Means and In-Betweens, Part 1: "Raisins and More Raisins," pages 13-19. The Used Numbers series was created by Technical Education Research Center (TERC), Lesley College, and the Consortium for Mathematics and Its Applications (COMAP) and is available from Dale Seymour Publications.

*Mathematical concepts and procedures are linked to our daily lives. These two lessons help create a classroom atmosphere where activities can extend over several days and connect with other mathematical strands, other disciplines, and the world outside the classroom.*

# Minnesota Mathematics and Science Teaching Project (MINNEMAST)

## Observing Temperature Changes

"Observing Temperature Changes" provides students with an in-depth learning experience which integrates mathematics and science.

In this lesson, students link conceptual and procedural knowledge as they record and graph temperature fluctuations over time.

Students begin the lesson by learning how to hold and read a thermometer. They then place several thermometers around the school both indoors and outdoors. Next, a team of students records the temperatures on the hour every school hour for several days. They learn to graph the recorded data from this scientific experiment and deepen their understanding of the connections between mathematics and science.

Topics for classroom discussion include: comparisons of the warmest and coolest parts of the day, comparisons of the warmest and coolest locations, and reasons for these temperature differences.

- For further information about MINNEMAST turn to pages 21 and 105.

From MINNEMAST Unit 20, Using Larger Numbers, Section 3, Lesson 16: "Observing Temperature Changes," pages 82-87. MINNEMAST materials were developed by the Minnesota Mathematics and Science Teaching Center at the University of Minnesota. For further information contact The Learning Team.

---

### Lesson 16: OBSERVING TEMPERATURE CHANGES

In this lesson, the children make actual thermometer readings and record temperatures on tables and graphs. They study temperature variations throughout a day and begin recording data on their Daily Weather Records. This lesson will probably take 4 days.

MATERIALS

- 14 thermometers
- red crayons, 1 per child
- masking tape
- Worksheets 49 through 59

PREPARATION

Thermometers should be placed outdoors on the north, east, south and west walls of the school building. Use masking tape. If you cannot find any place to which the tape will stick, tape each thermometer to the top of a large carton. Before you conduct Activity B, have four reliable children place these boxes outdoors, alongside the appropriate wall. Show them which is the north (etc.) wall the first time so they can find it by themselves next time. Tape a fifth thermometer in a corridor at about the average eye level of your class.

PROCEDURE

Activity A

Give a thermometer to each group of three or four children. Ask each child to write on scratch paper the temperature reading shown on his group's thermometer. The readings made by different groups may vary one or two degrees because of differences in the thermometers due to manufacturing methods. Show the children how to hold the thermometer for reading so that the top of the red column is at eye level and directly in front of them. Have them try holding the thermometer higher and lower than eye level and to one side, in order to see the distortions that can occur.

**The *Standards* state that in grades K– 4 the study of mathematics should include opportunities to make connections so that students can–**

- Link conceptual and procedural knowledge;

- Relate various representations of concepts or procedures to one another;

- Recognize relationships among different topics in mathematics;

- Use mathematics in other curriculum areas;

- Use mathematics in their daily lives.

# University of Georgia Geometry and Measurement Project

### 3.14 Visualizing Cubic Units of Volume

SCHOOL

Predict _____
Count _____

HOUSE

Predict _____
Count _____

BANK

Predict _____
Count _____

HOSPITAL

Predict _____
Count _____

POST OFFICE

Predict _____
Count _____

OFFICE BUILDING

Predict _____
Count _____

## Visualizing Cubic Units of Volume

This lesson helps students relate various representations of volume to one another. The lesson therefore helps students develop an understanding of the connections between mathematics, architecture, and construction.

The goal of the lesson is for students to visualize cubic units of volume in solid figures and to begin to compare and measure volumes of solids made from cubic blocks. Each student is given a set of cubes, enough to construct a small building. They estimate and count the number of blocks that they use. Then, working in small groups, the students use the blocks to create buildings of different shapes. A discussion follows in which they compare the volumes of the buildings.

- For further information about the University of Georgia Geometry and Measurement Project turn to pages 18 and 116.

From the University of Georgia Geometry and Measurement Project, Strand 3, <u>Space & Volume; Temperature; Weight</u>, Lesson 14: "Visualizing Cubic Units of Volume." Project materials were created at, and are available from, the University of Georgia.

*Estimation skills enhance students' ability to deal with quantity in everyday situations. The lessons below explore estimation strategies and help students understand when it is appropriate to estimate, how close an estimate is required, and whether the estimation results are reasonable.*

## Mathematics Curriculum and Teaching Program (MCTP)

### More Than, Less Than

"More Than, Less Than" helps students determine the reasonableness of results.

In this lesson, students estimate the length of their classroom. A "bracketing" approach is used in which the class agrees upon a lower limit and upper limit for the length of the room – for instance, more than one foot and less than one mile. Students are encouraged to explain and share their estimates. As the class discusses and refines their estimates, more and more reasonable estimates are heard. Towards the end of the lesson, the group makes a final estimate. They then actually measure the room. The final estimate is often close to the actual measurement.

Additional in-class activities include estimating the number of pages in a book and the number of beans in a jar.

• For further information about MCTP turn to pages 16, 32, 51, 74 and 102.

From MCTP, <u>Activity Bank</u>, Volume I, Chapter 6,"More Than, Less Than," pages 281-284. Materials were assembled by MCTP and are available from the Curriculum Corporation and NCTM.

### More than, less than

*Children are asked to look at the length of a room and give upper and lower limits of a measurement. These are gradually narrowed to a range with which the whole class agrees.*

**Why we need to have good estimation skills**

WE OFTEN NEED TO ESTIMATE.
SOMETIMES OUR ESTIMATES ARE VERY PRECISE, OTHER TIMES ONLY ROUGH BOUNDS ARE GIVEN.
FOR INSTANCE, HOW LONG WOULD IT TAKE TO GET TO MY FRIEND'S HOUSE?
YOU MIGHT SAY, I'LL BE THERE BETWEEN ONE AND TWO O'CLOCK.
WILL I HAVE ENOUGH MONEY FOR THE THINGS I'VE CHOSEN ONCE I GET TO THE CHECKOUT?
YOU WILL ESTIMATE IT'LL COST BETWEEN $15 AND $20 AND SO ON

WE ARE GOING TO DEVELOP A NUMBER OF ESTIMATES OF THIS KIND, WHERE WE BRACKET OUR ESTIMATES, THAT IS, FIND AN AGREED RANGE.
IT MIGHT SEEM FAIRLY ROUGH, BUT YOU MAY BE SURPRISED HOW ACCURATE IT TURNS OUT TO BE

It is essential to produce a very small starting lower bound, in order for the activity to fulfill its potential. Trial schools have found that expressions like *'if your life depended on it'* help to convey the message that you are looking for an extreme lower bound. Other teachers have actually started the ball rolling: *'Well, it's obviously greater than one metre and less than a million metres.'*

This consensus builds up a 'team approach' to the problem.

**An easily-agreed-upon lower limit**

Identify a distance inside or outside the classroom. The length of the classroom side wall may be a good starting point.

CAN YOU GIVE ME A MEASUREMENT THAT IS CERTAINLY LESS THAN THE LENGTH OF THIS ROOM?

ONE METRE (giggle)

DOES EVERYONE AGREE THAT THE ROOM IS *DEFINITELY* LONGER THAN ONE METRE? (One hopes so!)

**An easily-agreed-upon upper limit**

BETTY, CAN YOU GIVE ME A MEASUREMENT THAT IS CERTAINLY MORE THAN THE ROOM'S LENGTH?

IS EVERYONE HAPPY WITH THAT? SO, EVERYONE AGREES THAT THE LENGTH OF THE CLASSROOM IS BETWEEN ONE AND ONE HUNDRED METRES

ONE HUNDRED METRES

*Emphasise that this is* absolutely *true, i.e. this answer is correct, even though we will 'come closer' in good time. Write the numbers on the chalkboard as shown above.*

**The *Standards* state that in grades K– 4 the curriculum should include estimation so that students can–**

• Explore estimation strategies;

• Recognize when an estimate is appropriate;

• Determine the reasonableness of results;

• Apply estimation in working with quantities, measurement, computation, and problem solving.

# Comprehensive School Mathematics Program (CSMP)

## Hand-Calculator Tug of War

"Hand-Calculator Tug of War" helps students to develop the skills of estimation, mental computation and problem solving.

Students play a numerical estimation game based on calculator use. The class is divided into two teams and each team is assigned a number on the number line; for example, Team A might be assigned 167 and Team B might be assigned 835. Team A may only add a number for its move, and Team B may only subtract. The teams alternate moves, and the first team to meet or pass the other team's number loses. By varying the numbers assigned to each team, teachers can help students improve their number sense.

• For further information about CSMP turn to pages 23, 26, 35, 38 and 91.

---

```
ACTIVITY H2:      HAND-CALCULATOR TUG OF WAR

PREREQUISITE:     None

OBJECTIVE:        Students will practice estimation and mental arithmetic.
```

Provide each student with a hand-calculator. Allow the students a few minutes to play with and explore the use of the calculator.

Divide the group of students into two teams (A and B) and draw this number line on the board.

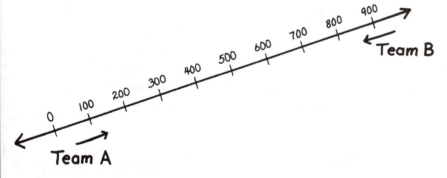

T: The students on Team A will start at 167 and the students on Team B at 835.

Let a student from each team mark the approximate location of their team's number on the number line. It might be helpful if the teams use different colors to locate their numbers.

T: The students on Team A can choose only ⊞ [some number], and the students on Team B only ⊟ [some number]. The teams take turns. The first team to meet the other team's number or to pass it will be the losing team.

T: Each time you make a play, you mark the approximate location of your team's new number on the number line.

From CSMP, Activities for TOPS, Strand III, Activity H2: "Hand-Calculator Tug of War," pages 3-4. CSMP materials were created by, and are available from, McREL Publications.

*Students need to understand numbers and develop number sense if they are to be successful in a society which increasingly quantifies, measures, identifies, collects, and names. They need to interpret the multiple uses of numbers which they encounter. These lessons offer hands-on approaches to understanding our numeration system and place-value concepts.*

## Curriculum Development Associates Math (CDA)

### Ten-Sticks and Loose Beans

"Ten-Sticks and Loose Beans" uses physical materials to relate counting, grouping, and place-value concepts.

Loose beans and sticks are used to model computation algorithms. The page shown here provides practice in addition using numbers between 5 and 9. Each stick is shown with 5 beans grouped on the left side and the rest on the right. This is done so that a number can be quickly identified as 5, plus some more. The students learn to count from 5 on. When beans on both sides of two sticks are to be counted, the two 5's can be considered as 10 and students can count from 10 on. For example, students are asked: How many beans on the top stick? 5. How many beans on the other stick? 5 and 2 more is 7. How many beans on both sticks? 5 and 5 are 10 and 2 more is 12.

- For further information about Curriculum Development Associates Math turn to page 93.

From Curriculum Development Associates Math, <u>Drill and Practice</u>, "Ten-Sticks and Loose Beans," Part II, pages 110A-110D. Materials were developed by and are available from Curriculum Development Associates, Inc.

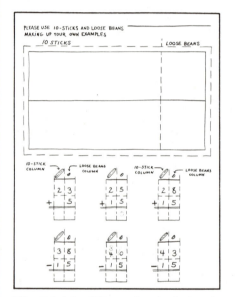

This activity helps children see how the algorisms for "carrying" and "borrowing" (or "regrouping") emerge as a record of experiments.

Assume that we have passed a rule that we cannot use more than nine loose beans. The activity might proceed as follows: "Please get me 23 beans, and put them in the top space."

| 10-sticks | loose beans |
|---|---|
| | |
| | |

"Now get me five more beans and put them in the lower space."

| 10-sticks | loose beans |
|---|---|
| | |
| | |

"How many beans are there all together?" There are 8 loose beans, which we record in the "loose beans" column, and two 10-sticks, which we record in the 10-stick column as follows:

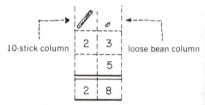

10-stick column      loose bean column

There is a total of 28 beans.

Now, consider an example like

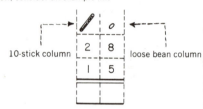

10-stick column      loose bean column

"Please get me 28 beans. Put them in the top space."

| 10-sticks | loose beans |
|---|---|
| | |

"Now get me 15 beans, and put them in the bottom space."

| 10-sticks | loose beans |
|---|---|
| | |
| | |

"How many are there all together?" There are 13 loose beans. But we said no more than nine loose beans could be used, so we can trade — ten loose beans for one stick, and record the trade by putting a small "1" in the 10-stick column.

one more 10-stick from the trade

(continued on page 110B)

The *Standards* state that in grades K–4 the mathematics curriculum should include whole number concepts and skills so that students can–

- Construct number meanings through real-world experiences and the use of physical materials;

- Understand our numeration system by relating counting, grouping and place-value concepts;

- Develop number sense;

- Interpret the multiple uses of numbers encountered in the real world.

# Nuffield Mathematics Teaching Project

To give children an idea of why 'rounding off' large numbers is both sensible and meaningful, we can suggest 'Numbers in Space' as a topic of interest. The following will probably provide suitable material to introduce this :

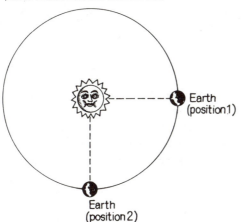

Earth is one of a family or set of 9 planets revolving or orbiting around the Sun. The nearest planet to the Sun is **Mercury**, then Venus, then Earth. The paths traced by these planets are not circular, but oval or elliptical.

Because the Sun is not in the centre of a circle, the distance between the Sun and each planet changes every second. So, we speak of distances between the Sun and the planets in terms of average or mean distances.

The average distance between the Earth and Sun is 92,912,000 miles. For Mercury it is 36,059,000 and for Venus 67,206,000 miles.

**1** These distances have been rounded off to the nearest **million, thousand** or **hundred** miles. State which you think it is.
**2** Round off these distances to the nearest million miles.

Here are some more average distances from the Sun to the other planets of the solar system :
| | |
|---|---|
| **Mars** | **141,568,700 miles** |
| **Uranus** | **1,782,227,700 miles** |
| **Neptune** | **2,792,717,000 miles** |
| **Jupiter** | **483,401,300 miles** |
| **Saturn** | **886,270,100 miles** |
| **Pluto** | **3,671,660,400 miles** |

A 'table' can be made in which the 9 planets of the solar system are listed to show their average distances from the Sun, in order. Begin with the planet nearest the Sun and give the distances to the nearest million miles.

| Name of planet | Average distance from Sun (in millions of miles) |
|---|---|
| Mercury | 36 |

The distance of the Moon from the Earth, at a particular time, is 238,857 miles. This could be written as :
**238·857 thousand miles.**

Here we have simply changed the unit of length from one mile to one thousand miles.

The Earth's diameter is 7913 miles. How could we write this in thousands of miles?

## Large Numbers and Indices

"Large Number and Indices" is filled with activities that use hands-on materials and real-world examples to help students understand numbers and place value.

These activities include:

- Using graph paper to illustrate the concept of a million. (How many 1/10-inch squares in a 1-inch square? In 10 1-inch squares? In 100 1-inch squares? etc.).

- Using peg boards to create a large number machine;

- Relating large numbers to populations, the solar system, and time;

- Using Base-10 Blocks to illustrate indices and expanded notation.

These activities provide students with experiences which will help them to understand and read large numbers.

- For further information about Nuffield Mathematics Teaching Project turn to pages 15 and 106.

From Nuffield Mathematics Teaching Project, Computation and Structure, Book 4, Lesson 5: "Large Numbers and Indices," pages 50-54, 57-63. Nuffield materials were developed at the Centre for Science Education and are available from Nuffield-Chelsea Curriculum Trust.

*Recognizing conditions that indicate the need for whole number operations is central to mathematics instruction. The lessons below help relate mathematical language to problem and real-world situations and develop a student's understanding of the fundamental mathematical operations.*

## Madison Project

### Can You Add and Subtract?

"Can You Add and Subtract?" helps students relate the mathematical language of operations to problem situations and informal language.

Students deepen their understanding of addition and subtraction by imagining that they are operating a pet store. On a typical day customers come and go, buying and returning various pets and supplies. Using addition and subtraction, students keep track of how much money is in the cash register. Dogs, cats and rattlesnakes are some of the animals that students might include in the activity.

- For further information about the Madison Project turn to pages 5, 46, 73, 85 and 101.

From the Madison Project, <u>Discovery in Mathematics</u>, Chapter 2: "Can You Add and Subtract?" pages 35-37. <u>Discovery in Mathematics</u> was written by the Madison Project and is available from Cuisenaire Company of America.

---

 **Chapter** 2
CAN YOU ADD AND SUBTRACT?

**[page 3]**

Suppose we are operating a pet store, and we come into the store on Monday morning. We unlock the door, unlock the cash register (which has quite a lot of money in it), and feed the animals.

Now, somebody comes in to buy something.

(1) What does he buy?

(2) How much does he pay us for it?

(3) Is there now **more** money in the cash register than there was when we opened up this morning, or is there **less**?

(4) How much less, or how much more?

(5) Now somebody comes in to return something. What does he return?

(6) How much money do we give back?

(7) Is there now more money in the cash register than there was when we first opened up this morning, or is there less?

(8) Do you know how much more, or how much less?

The *Standards* state that in grades K–4 the mathematics curriculum should include concepts of addition, subtraction, multiplication, and division of whole numbers so that students can–

- Develop meaning for the operations by modeling and discussing a rich variety of problem situations;

- Relate the mathematical language and symbolism of operations to problem situations and informal language;

- Recognize that a wide variety of problem structures can be represented by a single operation;

- Develop operation sense.

# Nuffield Mathematics Teaching Project

## 4   The operation of addition

In *Mathematics Begins* ❶ we saw that children sort and classify and partition sets of objects into sub-sets. They then also learn to count the number of elements in a set.

A number tells us how many elements a set has. For instance, all sets of objects whose elements can be matched with this set:

have the common property of *threeness*.

The *union* of two sets is the set of elements which belong to at least one of them.

For example, let set A consist of a bird, the cat Copper and the dog Bonzo:

and let set B consist of a biscuit, a ball and the cat Copper:

Then the union, written A ∪ B, consists of

This is the set of elements which belong to A or B or both.

The *intersection* of two sets consists of the set of elements which belong to both.

In this example, the intersection of A and B (written A ∩ B) consists of the cat Copper who is the only 'element' belonging to both A and B.

## The Operation of Addition

On this page from the teacher's text, teachers read about how students can be engaged in a variety of problem situations that would help them develop an understanding of addition.

Students explore different ways of reaching a target sum using two addends;

they look for patterns in their work; they observe the role of zero in addition; they complete addition problems that are missing sums or addends and, using a balance as an "equalizer," they look for equalities and inequalities.  In the example above, a set model of addition, students learn to identify common elements of sets.

Number strips and other aids provide visual representations of addition.  Further activities include puzzles, magic squares,

and games that develop proficiency with basic addition facts.

- For further information about Nuffield Mathematics Teaching Project turn to pages 13 and 106.

From Nuffield Mathematics Teaching Project, Computation and Structure, Book 2, Section 4: "The Operation of Addition," pages 58-69. Nuffield materials were developed at the Centre for Science Education and are available from Nuffield-Chelsea Curriculum Trust.

*There are a variety of computation techniques, including the use of calculators. Students need to learn how to select a computation method and determine whether their results are reasonable. These two lessons help to develop a student's computational skills, mathematical insights, and confidence in using various mathematical procedures.*

## Mathematics Curriculum and Teaching Program (MCTP)

### How Did You Do It?

"How Did You Do It?" helps students learn a variety of mental arithmetic techniques and encourages them to use mental arithmetic on occasions when they might otherwise have reached for a pen and paper.

The teacher begins the lesson by writing an addition, subtraction, multiplication or division problem on the chalkboard. He/she then asks the class to use mental arithmetic to determine the answer. After the group has agreed on the correct answer, the teacher asks for volunteers to explain their methods.

Students and teachers are frequently surprised by the variety and sophistication of the methods evoked by this process. Students are able to watch themselves and each other think. They often find that they can use a variety of approaches when pen and paper are not available or necessary.

- For further information about MCTP turn to pages 10, 32, 51, 74 and 102.

From MCTP, Activity Bank, Volume I, Chapter 6: "How Did You Do It?" pages 273-276. Materials were assembled by MCTP and are available from the Curriculum Corporation and NCTM.

### How did you do it?

*The teacher gives a mental arithmetic problem and the answer is quickly established. Then children are asked to describe the strategies they used. The ensuing discussion tends to give children confidence to use mental arithmetic and expand their repertoire of strategies.*

The teacher will need to consider carefully the level of difficulty. Children will be attempting to solve it mentally. Some possibilities might be 37 + 15 (adding and subtracting two-digit numbers), or 23 x 4 (multiply or divide by a one-digit number), or 1 + 2 + 3 + 4 + 5 + 6 + 7 + 8 + 9 + 10 (or variations of these).

Teachers report that it is important to establish the answer first, (the correct answer is only a side issue) otherwise pupils will be waiting for it. Also, if the first few pupils to answer have impressive methods, but are wrong, the activity can fall flat. By being fairly casual about the answer ('So let's just have the answer..... fine, now then let's ......'), pupils can then really focus on the processes involved.

Write the problems <u>across</u> the board, rather than down. Trial schools report that this is more likely to encourage pupils to move from standard algorithmic approaches.

**Write a problem on the chalkboard.**

The problem need not be particularly creative, the interest lies in the methods of solution, which often turn out to be surprisingly creative. Avoid giving calculations which are too complicated — the purpose is not to see if they can do it, but **how** they tackle it. However, don't make the problem so simple that the answer becomes automatic. The example shown below is suitable for year 3.

Having agreed on the correct answer first ('*Everyone say what the answer is*'), ask for volunteers to explain their methods.

As the pupils relate their methods, try to summarise these on the board. Teachers need to be careful not to 'rate' these methods. The criteria for deciding whether a method is OK are:

- Does it work?
- Is the pupil happy with it?

TODAY I'M NOT AS INTERESTED IN THE ANSWER — WHAT IS IT BY THE WAY?... 71. OKAY, BUT RATHER IN ALL THE DIFFERENT WAYS YOU DID IT IN YOUR HEADS

46 + 25 =

The *Standards* state that in grades K–4 the mathematics curriculum should develop whole number computation so that students can–

- Model, explain, and develop reasonable proficiency with basic facts and algorithms;

- Use a variety of mental computation and estimation techniques;

- Use calculators in appropriate computational situations;

- Select and use computation techniques appropriate to specific problems and determine whether the results are reasonable.

# Lane County Mathematics Project

## Calculators and Cards

CALCULATORS AND CARDS

1. Use any two numbers on the cards. Complete each addition problem below.
   Check your work with a calculator.

   `97`  `204`  `297`  `405`  `498`  `607`

   a. ☐ + ☐ = 301

   b. ☐ + ☐ = 1012

   c. ☐ + ☐ = 501

   d. ☐ + ☐ = 1105

   e. ☐ + ☐ = 595

   f. ☐ + ☐ = 795

   g. ☐ + ☐ = 394

   h. ☐ + ☐ = 811

   i. ☐ + ☐ = 609

   j. ☐ + ☐ = 903

   k. ☐ + ☐ = 702

   or

   ☐ + ☐ = 702

2. Now use any three numbers on the cards. Complete these problems.

   a. ☐ + ☐ + ☐ = 598

   b. ☐ + ☐ + ☐ = 1202

   c. ☐ + ☐ + ☐ = 1402

   d. ☐ + ☐ + ☐ = 1510

   e. ☐ + ☐ + ☐ = 1309

   f. ☐ + ☐ + ☐ = 1107

"Calculators and Cards" provides activities which help students explore mental arithmetic and estimation techniques as well as the use of a calculator.

Students are given a worksheet containing several addition problems and are asked to select two numbers (one per "card") to solve each problem. To accomplish this, students mentally add two-and three-digit numbers and use place-value concepts to make reasonable estimates.

Students then check their work with a calculator.

- For further information about the Lane County Mathematics Project turn to pages 6, 66 and 99.

From the Lane County Mathematics Project, <u>Problem Solving in Mathematics</u>, Grade 4, Section III: "Calculators and Cards," pages 91-92. The <u>Problem Solving in Mathematics</u> series was created by the Lane Education Service District and is available from Dale Seymour Publications.

*Insight and intuition about two-and three-dimensional shapes, the interrelationships of shapes, and the effects of changes in shapes are important aspects of spatial sense. These lessons encourage students to investigate, experiment, and explore with everyday objects in order to develop their geometric knowledge.*

# University of Georgia Geometry and Measurement Project

## Building Shapes from Small Squares

In this lesson, students investigate the results of combining and re-arranging square tiles and cubes. These activities help students to develop an understanding of shape.

Using tiles and grid paper, students create shapes that can be formed from two small squares. Students find all possible arrangements of the two squares. They test their results by placing the shapes on top of one another and comparing them directly. Students repeat the exercise using three tiles, four tiles, and cubes, and then discuss their results.

As students model, classify, and describe these shapes, they gain a greater understanding of geometry and spatial relations.

- For further information about the University of Georgia Geometry and Measurement Project turn to pages 9 and 116.

From the University of Georgia Geometry and Measurement Project, Strand 2, Angle, Surface and Area, Lesson 49: "Building Small Shapes From Squares," pages 1-3. Project materials were developed at, and are available from, the University of Georgia.

---

### 2.49 Building Shapes from Small Squares

Teacher _____     Grade _____ Date _____

**Level: 2-4**

**Purpose:** Investigating shapes that can be formed from small squares

**Materials:** One-inch square tiles, grid paper with one-inch squares (see attached pattern)

**Activities:**

(1) Give students some square tiles and grid paper. Have students find all possible arrangements of two squares. Have the children record their shapes on the grid paper.

After the children have found what they think are all of the arrangements, ask them to mark the ones that have sides that are in complete contact with each other (see the first example below).

| Complete contact | Not complete contact | Not complete contact |

Have each child tape two squares together. Have them fit this new shape on each of the arrangements that they marked. There is only one arrangement for two squares although the different positions for the shape may make it appear different to the children. For example, the following look different, but one will fit on top of the other.

The *Standards* state that in grades K– 4 the mathematics curriculum should include two and three-dimensional geometry so that students can–

- Describe, model, draw and classify shapes;

- Investigate and predict the results of combining, subdividing and changing shapes;

- Develop spatial sense;

- Relate geometric ideas to number and measurement ideas;

- Recognize and appreciate geometry in their world.

# Developing Mathematical Processes (DMP)

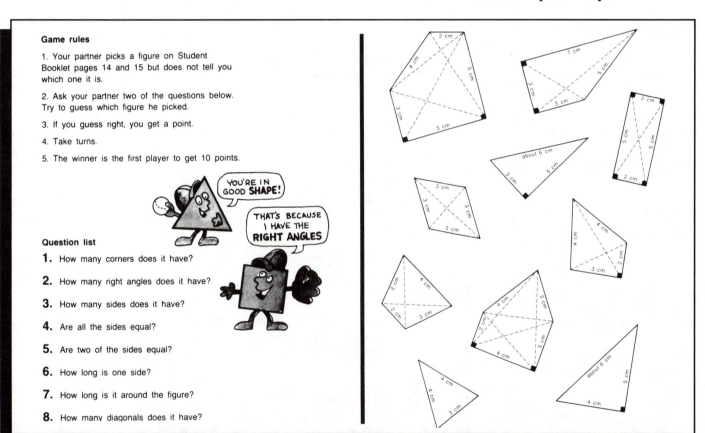

**Game rules**

1. Your partner picks a figure on Student Booklet pages 14 and 15 but does not tell you which one it is.

2. Ask your partner two of the questions below. Try to guess which figure he picked.

3. If you guess right, you get a point.

4. Take turns.

5. The winner is the first player to get 10 points.

YOU'RE IN GOOD **SHAPE!**

THAT'S BECAUSE I HAVE THE **RIGHT ANGLES**

**Question list**

**1.** How many corners does it have?

**2.** How many right angles does it have?

**3.** How many sides does it have?

**4.** Are all the sides equal?

**5.** Are two of the sides equal?

**6.** How long is one side?

**7.** How long is it around the figure?

**8.** How many diagonals does it have?

# Geometric Figures

"Geometric Figures" develops students' ability to describe and classify two-dimensional figures of various shapes and sizes.

In Part 1 of this lesson, students play "Figuroni," a game in which they describe and classify rectangles, triangles, and pyramids of varying proportion.

In Part 2, illustrated here, students working in teams of two are given several convex shapes and are encouraged to consider questions such as: What things do I have to tell my friend so she can make a figure like mine? What do I have to tell her so she knows which figure I mean? This activity helps students talk about math and deepen their understanding of geometric shapes.

In Part 3, students focus their attention on certain combinations of attributes by

marking figures with four equal sides, a figure with only one right angle, and others. Activities that students may do as a whole class are also provided.

- For further information about DMP turn to pages 20,40, 43, 45 and 94.

From DMP, Book 48, Activity 48H: "Geometric Figures," pages 37- 39. DMP was developed at the University of Wisconsin-Madison and is published by Delta Education.

*Students can grasp the usefulness of mathematics in the everyday world through the development of their understanding of measurement. Measurement is also a natural context in which to introduce fractions and decimals. These lessons provide children with experience in developing the vocabulary and skills associated with measurement.*

# Developing Mathematical Processes (DMP)

## Measuring

"Measuring" deepens students' understanding of various types of units of measurement.

In this activity, part of a larger unit on measurement, students discover that using the sense of touch to measure temperature is unreliable when temperatures vary only slightly. In Part 1 of the activity, students realize the need for an instrument to measure temperature. They make a thermometer and conduct experiments in which they practice reading it.

In Part 2, students conduct several short experiments in which they use a thermometer scaled in Fahrenheit degrees.

In Part 3, pictured here, students are introduced to the thermometer scaled in Celsius degrees. Students are given a chart indicating the average temperatures on the Celsius scale for eight cities. They are then asked to find the greatest and least differences between temperatures.

- For further information about DMP turn to pages 19, 40, 43, 45 and 94.

From DMP, Book 63, Activity 63A: "Measuring" pages 9-14. DMP was developed at the University of Wisconsin-Madison and is published by Delta Education.

### Temperatures in the United States

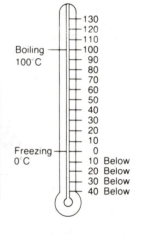

| City | Average Temperature* January °C | Average Temperature* July °C |
|---|---|---|
| Chicago | 3° below 0° | 24° |
| St. Louis | 0° | 25° |
| San Francisco | 11° | 15° |
| Minneapolis | 11° below 0° | 22° |
| Miami | 19° | 28° |
| Dallas | 8° | 29° |
| Juneau | 4° below 0° | 13° |
| Honolulu | 22° | 26° |

*Temperatures are in degrees Celsius.

**1.** Would you rather swim in water having a temperature of 10°C or 30°C? Which is warmer?

**2.** Would a marble fall faster through water at 10°C or at −10°C (below 0)?

**3.** Which city is
warmest in January?
warmest in July?
coldest in January?
coldest in July?

**4.** Which city has the greatest difference between July and January temperatures? how many degrees difference?

**5.** Which city has the smallest difference between July and January temperatures? how many degrees difference?

**6.** The record high temperature in St. Louis is 46°C. How much hotter is that than the average July temperature?

**7.** In Chicago the record low is 31°C below 0°C. Is this hotter or colder than the average January temperature?

**8.** San Francisco's record high temperature is 41°C. How much hotter is that than the average July temperature? How much hotter is that than the average January temperature?

The *Standards* state that in grades K–4 the mathematics curriculum should include measurement so that students can–

• Understand the attributes of length, capacity, weight, area, volume, time, temperature and angle;

• Develop the process of measuring and concepts

related to units of measurement;

• Make and use estimates of measurement;

• Make and use measurements in problem and everyday situations.

# Minnesota Mathematics and Science Teaching Project (MINNEMAST)

Lesson 6: SCALING WITH AN ARCHITECTS RULER

An architects ruler is made up of six different scales, in various multiples of a 1 → 1 scale. To make a representation of a rectangle in a scale of 1 → 3, you measure the original with the 1 → 1 scale and then draw the representation, using the same number of units on the 1 → 3 scale.

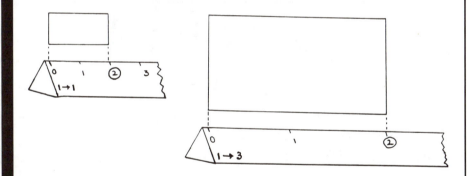

With this handy tool you can increase the dimensions of the original shape by a factor of 3, without using grid paper, counting units, or solving any multiplication equations.

The children are provided a printed sheet from which they make their own rulers. They first learn to use them, and then find out why they work. When they use their own standard units to mark off scales along blank lines, it helps them to see how one scale is a multiple of another. The architects ruler is another form of the parallel number lines the children were introduced to in Unit 17, where one line represented the number of kangaroo jumps and the other represented the number of units in each jump.

Prepare an architects ruler from the printed construction sheet facing page 42 before you teach this lesson.

## Scaling with an Architect's Ruler

"Scaling with an Architect's Ruler" illustrates how measurement is an essential element of an architect's trade.

In this lesson, students create an architect's ruler using strips of construction paper, scissors, and a worksheet. With the ruler, students can scale the dimensions of various shapes up and down. They learn how to use the ruler and why it works. Additional activities are suggested in which students create rulers with a one-to-four scale, a one-to-five scale, and a one-to-six scale. These rulers are also used to scale students' drawings up and down.

• For further information about MINNEMAST turn to pages 8 and 105.

From MINNEMAST, Unit 18, Lesson 6: "Scaling with an Architect's Ruler," pages 41-46. MINNEMAST was developed by the University of Minnesota Mathematics and Science Center. For further information contact The Learning Team.

# K-4 Standard 11: Statistics and Probability

*Students' questions about the world around them can often be answered in ways leading to greater understanding of statistics and probability. These lessons provide students with opportunities to collect, organize, describe, display and interpret data as well as make decisions and predictions on the basis of that information.*

## Teaching Integrated Mathematics and Science (TIMS)

### Plant Growth

"Plant Growth," like all of the TIMS units, integrates mathematics and science. This experiment provides an opportunity to integrate the study of measurement, statistics, and probability with an initial look at photosynthesis.

Students sprout seeds and measure the height of the sprouts over a three-week period. They use data tables to record dates, the number of days since their last measurement, and the height of their plants. They also record their own predictions as to the outcome of each measurement. Once the data has been collected, recorded and described, students graph and analyze the results.

Students also gain experience with variables by varying the amount of soil, light and water and seeing how these variables affect the height of their plants. The Student Lab Write-Up page shown at right is one of the probability activities that students work on after their seeds have sprouted. In this activity the amount of soil is the variable.

• For further information about TIMS turn to pages 2, 47, 54 and 114.

From TIMS, Experiment Unit 206, "Plant Growth." The TIMS project materials were developed at, and are available from, the University of Illinois at Chicago.

---

Student Lab Write-Up        5        206.3 Plant Growth

13. In Mr. Jones' class the children collected data for the frequency distribution shown below.

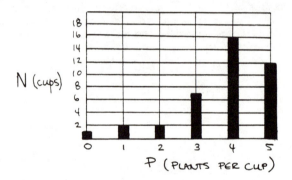

How many cups were planted?_____

What is the probability of no plants sprouting?_____

What is the probability of all the plants sprouting?_____

14. Billy wants to see if the volume of the dirt in the cup changes how tall a plant grows. He does two experiments shown below.

| Experiment 1 | Experiment 2 |

Will these experiments show how the volume of dirt will affect the height of the plant?_____Explain. _____

_____

The *Standards* state that in grades K–4 the mathematics curriculum should include experiences with data analysis and probability so that students can–

- Collect, organize and describe data;

- Construct, read and interpret displays of data;

- Formulate and solve problems that involve collecting and analyzing data;

- Explore concepts of chance.

# Comprehensive School Mathematics Program (CSMP)

L12     Two Dice #1

CAPSULE LESSON SUMMARY

Examine a die noting that it is a cube and what is on the six faces of the cube. Discuss the chances of getting various outcomes when tossing one die. Determine all the possible outcomes when tossing two dice and picture them in a coordinate graph. Discuss the possible sums (add the numbers obtained on the two dice) you could get when tossing two dice and find that some sums have a better chance of occurring than others.

MATERIALS

Teacher:  Two dice; colored chalk; die cutout
Student:  None

NOTE: Before the lesson begins, make a paper cutout of a die as illustrated below. Make the squares between 5 cm and 6 cm on a side. The dotted lines indicate fold lines.

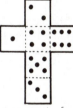

The dice provided in your classroom set of materials are red and white; however, any two different colors could be used.

## Two Dice #1

In "Two Dice #1" students learn to construct, read and interpret displays of data as well as formulate and solve problems that involve the collecting and analyzing of data.

In this lesson on theoretical probability, students examine a die and discuss its shape and what is on its six faces. They discuss the chances of getting various outcomes when tossing one die. When a second die is introduced, the class develops all possible outcomes that might occur from tossing two dice by representing them on a grid or coordinate graph. In analyzing their graph, they discuss the possible sums that might result from tossing two dice and conclude which sum has the greatest chance of occurring.

- For further information about CSMP turn to pages 11, 26, 35, 38 and 91.

From CSMP, <u>Mathematics for the Upper Primary Grades, Part II</u>, Activity L12: "Two Dice #1," pages 69-76. CSMP materials were created by, and are available from, McREL Publications.

# K-4 Standard 12: Fractions and Decimals

*An understanding of fractions and decimals expands students' appreciation of the usefulness and power of numbers and number systems. The lessons below help develop an awareness of how fractions and decimals are used in everyday life and serve as crucial extensions of students' knowledge about numbers.*

## Journeys in Mathematics

### Comparing Breakfast Cereals

"Comparing the Costs of Breakfast Cereals," an activity that all students will be able to relate to, provides students with an opportunity to apply their knowledge of fractions and decimals.

After collecting 6-ounce cereal boxes, students work with fractions and decimals to discuss and graph:

- Cereal ingredients -- What percent of each cereal is sugar?
- Box size – Is Box A  1/3 larger then Box B?
- Cost – Is box B 1/2 the price of Box C?
- Taste – Why is Box C sweeter then Box D?
- Net weight – How does the box size compare to the net weight of the contents?

- For further information about Journeys in Mathematics turn to pages 4, 49 and 98.

From Journeys in Mathematics, <u>Data and Decisions</u>: "Comparing the Costs of Breakfast Cereals," Section 2, pages 57- 80.  Developed by Education Development Center and published by WINGS for learning.

## 2 Comparing Breakfast Cereals

Students compare breakfast cereals, collecting data on taste, cost, and other factors.  They organize and analyze the information for use in Section 3 when they write consumer reports and create advertisements recommending the cereals to particular groups of consumers, such as children, parents, and nutritionists.

In Activity 7, students consider *what matters* when deciding which cereals to recommend.  They collect information from the cereal boxes and record the information in a table.  In Activity 8, the class conducts a poll to determine which cereal students in the class like best.  They then display the data in a bar graph.  Students also conduct a poll to determine which cereals adults like best.  Then, in Activity 9, students construct double bar graphs to compare the student taste preference data to the adult data.  In Activity 10, students calculate and compare the costs per ounce of the cereals and consider whether the most expensive cereals are the best-tasting.  Finally, in Activity 11, they collect data about other aspects of cereals — for example, amount of fruit or prizes — and present their data to the class.

### Objectives

This section both extends the objectives listed in Section 1, and covers the additional objectives below.

Students will:

1. Plan and execute a strategy for collecting and analyzing data, including information from
   - polls
   - product labels

2. Construct double bar graphs to present a comparison of two data groups.

3. Calculate and use price per ounce data as part of a problem-solving strategy.

**The *Standards* state that in grades K–4 the mathematics curriculum should include fractions and decimals so that students can–**

- Develop concepts of fractions, mixed numbers and decimals;

- Develop number sense for fractions and decimals;

- Use models to relate fractions and decimals;

- Use models to explore operations on fractions and decimals;

- Apply fractions and decimals to problem situations.

# School Mathematics Study Group (SMSG)

## Rational Numbers Describe Points on the Number Line

"Rational Numbers Describe Points on the Number Line" helps students to develop number sense for fractions and decimals.

In this lesson, students review and extend the idea of rational numbers describing points on a number line. As seen on the left, these exercises provide practice labeling points on the number line, using both whole numbers and rational numbers. The students begin by locating the point halfway between 0 and 1 and using the rational number as a label. Finding other points where the segments are congruent to the one whose end points are marked by 0 and 1/2, they can then suggest rational numbers that describe those points.

- For further information about SMSG turn to pages 72 and 111.

VIII-5.  <u>Rational numbers describe points of the number line</u>

<u>Objective</u>:     To review and extend the idea of rational numbers describing points on the number line.

<u>Vocabulary</u>:    (No new words.)

<u>Materials</u>:     Line on chalkboard with 5 equally spaced points (1 foot apart).

<u>Suggested Procedure</u>:

Label the left point on the line the 0-point.  Then have a child label the other marked points by writing 1, 2, 3, below the line.  Now, let us use the rational numbers to label points on the line.  What point can be labeled by 1/2 ?  Guide children by suggesting the separation of the first unit segment into two congruent parts.

Now let us find other points where the segments are congruent to the one whose end points are marked by 0 and 1/2.  Mark these points on the number line.

Then have children suggest rational numbers that will describe these points.  The line may appear as follows:

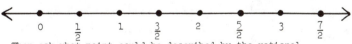

Then ask what point could be described by the rational number 1/4.  Again, observe that there is a smaller segment.  Again, lay off segments equivalent to this point.  Identify the rational numbers that describe these points, such as,

<u>Pupil's Book</u>, <u>pages</u> 434 and 435:  Children describe points of number line using both the whole numbers and the rational numbers.

From SMSG, <u>Mathematics for the Elementary School</u>, Book 3, Part II: "Rational Numbers Describe Points of the Number Line," Section VIII, pages 703-05.  SMSG was developed at the School of Education at Stanford University.  For further information contact The Learning Team.

*Observing and describing regularities in events, shapes, designs and sets of numbers is a fundamental aspect of mathematics and serves as a foundation for more abstract ideas. These lessons encourage students to look for and create patterns and mathematical relationships which will add to their general mathematics knowledge and help prepare them for more advanced mathematical work.*

## Comprehensive School Mathematics Program (CSMP)

### Taxi Geometry

"Taxi Geometry" provides an opportunity for students to practice representing and describing mathematical relationships.

In "Taxi Geometry" students work on a grid to represent and find lengths of various paths, using the rules of taxi geometry. In taxi geometry, the distance between two points on a grid is defined as the sum of the numbers of horizontal and vertical units between the two points (analogous to the number of city blocks between two sites represented on a city map). Within this measurement world, students find the distance between various points and identify geometric patterns associated with the locus of points reached when given a distance from a fixed point on the grid.

• For further information about CSMP turn to pages 11, 23, 35, 38 and 91.

From CSMP, <u>Mathematics for the Upper Primary Grades</u> Part II, Activity G3: "Taxi Geometry," pages 12-29. CSMP materials were created by, and are available from, McREL Publications.

T:      This is the map of the city where Nora lives. Here is Nora's house (point to "N"), and here is her grandmother's house (point to "G"), and here is Nora's school (point to "S"), and here is the library (point to "L"). Who can tell us something about our friend Nora?

Let students briefly recall some of the previous activities involving Nora. Encourage them to discuss long and short walks and counting the number of blocks in a path. If possible during this discussion, ask students to guess which place, L, S or G, is closest to Nora's house.

T:      Suppose it is raining and Nora wants to take a shortest path from her house (N) to school (S). How many blocks is a shortest path from N to S? Who can show us a way for Nora to walk?

Call on a volunteer to trace a path from N to S. Ask the class if anyone can find a shorter path. When the students agree on a shortest path, draw it on the grid board. Ask a student to count the blocks in this shortest path. (Answer: 15 blocks)

Repeat the questioning to get a shortest path from N to G, and a shortest path from N to L, and a shortest path from S to L. Possible responses are illustrated below.

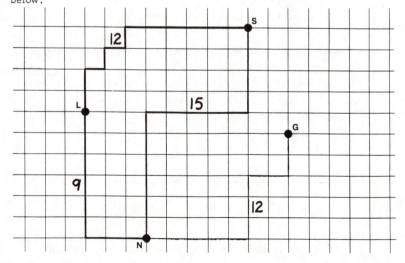

The *Standards* state that in grades K–4 the mathematics curriculum should include the study of patterns and relationships so that students can–

- Recognize, describe, extend, and create a wide variety of patterns;

- Represent and describe mathematical relationships;

- Explore the use of variables and open sentences to express relationships.

# University of Illinois Arithmetic Project (UIAP)

## Maneuvers on Lattices

"Maneuvers on Lattices," like all of the UIAP lessons, helps elementary school students pursue mathematics beyond the usual limits of elementary school. This lesson provides an opportunity for students to practice using open sentences to express mathematical relationships. It can also be used as an introductory lesson on functional relationships.

The lesson begins with the presentation of a table or "lattice" of natural numbers. Students move around the lattice according to directional arrows. They complete the unfinished table, using number patterns and addition to determine entries. The system is extended through class discussion and exploration.

Moves around the lattice become more complex with the use of multiple arrows. Students investigate special cases, such as moves at the boundaries of the lattice. Further extensions include lattices with fractional numbers, lattices with negative numbers, non-rectangular lattices, multiplicative lattices, and three-dimensional lattices.

- For further information about UIAP turn to pages 42 and 117.

From UIAP, <u>Maneuvers on Lattices: An Example of Intermediate Invention</u>, pages 1-15. The UIAP materials were developed at the Education Development Center, Inc. For further information contact The Learning Team.

---

## AN EXAMPLE OF "INTERMEDIATE INVENTION"

# MANEUVERS ON LATTICES

### BY DAVID A. PAGE

This "intermediate invention" has come about through the active cooperation of children ranging from first grade to high school. The author prefers not to estimate the fraction of the content here which is attributable to the collection of children in schools who have worked with it.

\* \* \*

As presented here, this topic begins in first grade. A teacher starting it for the first time with a fourth grade class would need to enter the topic by a faster, more omplex route.

The presentation here is necessarily continuous. In schools, on the other hand, it should be assigned a week here and there over the several years.

The student first coming to this topic, whatever his grade level, is assumed to be familiar with the physical interpretation of number-symbols (hereafter called numbers) such as 1, 2, 3, . . . , 100, . . . He should be able to get 17 blocks out of a box upon request. If he "sees and hears" whole numbers easily, it is not necessary that he can "make his figures well" at the outset.

### 1. First Grade

Put on the board, as pupils watch, the following table ("lattice") of numbers:

```
60  61  62
50  51  52  53  54  55  56  57  58  59
40  41  42  43  44  45  46  47  48  49
30  31  32  33  34  35  36  37  38  39
20  21  22  23  24  25  26  27  28  29
10  11  12  13  14  15  16  17  18  19
     1   2   3   4   5   6   7   8   9
```

After completing the first row or two, have students tell "what number comes next". Do this several times, and then complete the table by skipping around among the various unfilled places with students telling what number goes there. Include for example:

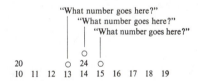

"What number goes here?"
"What number goes here?"
"What number goes here?"

```
           ○
20      ○ 24 ○
10  11  12  13  14  15  16  17  18  19
```

and several other examples where students tell what number goes directly to the right, left, above, or beneath a given number. Do not use many words at all. As a challenge, do it all in pantomime. Start the table, and after a couple of rows, hesitate and point to a spot and look puzzled instead of saying anything. (And so on.)

The table is complete to somewhere up above 50. (The more experienced and brighter the students, the less table needed on the board.)

"We are going to have a sort of secret code for writing numbers. It uses that table. Here is a number in code:

5↑

What number do you guess this stands for?"

In most of the classes in which this has been tried, a response in agreement with the teacher's plan has been forthcoming. It is important for the teacher to know and

Professor David A. Page, Director of the University of Illinois Arithmetic Project, joined Educational Services Incorporated in August, 1963, on leave from the University of Illinois. At ESI Professor Page is developing materials and techniques for instructing elementary school teachers in mathematics teaching.

# Section I  The *Standards* and Examples for Their Application

**Grade level
5—8**

*The problem-solving process helps students experience the power of mathematics. These lessons introduce students to a variety of problem-solving strategies and approaches. They require the development of genuine problem-solving skills which go beyond simple word problem exercises.*

## Regional Math Network (RMN)

### What's the Problem at the Boston Marathon?

In this lesson, students use problem-solving approaches to investigate and understand mathematical content.

Students begin by using a Press Kit (provided in a teacher packet) to research facts about the Boston Marathon. They discuss types of questions that could be asked about the marathon, such as: How many runners were there? What percent of the runners were women? How many ways did runners train for the run? Students then choose one or two pieces of data to formulate and solve their own word problems. They determine appropriate problem-solving strategies for each question that they devise, and share their questions and strategies with the class.

• For further information about RMN turn to pages 36 and 110.

From Regional Math Network, <u>Sports Shorts</u>, "Boston Marathon," pages 95-122. Regional Math Network materials were created at the Harvard Graduate School of Education and are distributed by Dale Seymour Publications.

---

## EDITOR'S NOTES

<u>TITLE:</u>  What's the Problem at the Boston Marathon

<u>NATURE OF ACTIVITY:</u>  Problem Posing

<u>OBJECTIVES:</u>  To formulate word problems
To identify relevant data in a problem solving situation
To choose appropriate problem solving strategies

<u>PRE-SKILLS:</u>  Some problem solving experience

<u>MATERIALS:</u>  Fact sheets (see Press Kit), brochures, flyers, etc. about facility (optional)

<u>NOTES:</u> This activity requires students to focus on the facts in a problem situation as well as on the question and solution. Discuss "fact" sources other than those in the teacher packet (Press Kit). Students may be able to bring in program books, brochures, etc. or have information from personal trips, T.V. viewing, etc. To start the thinking process, choose one or two pieces of data and have students brainstorm to produce possible questions. Focus on fluency first, then select those questions that can be solved with the students' background in mathematics. Discuss the kinds of questions that could be asked: How many? How much more? What fraction? What percent? How many ways? What's the least? Etc.

Encourage questions that require varied strategies such as single computation, multiple computations, sketching, efficient counting, chart or list, etc. Plain paper may be used so the number of problems is not limited to five. Students may fold over the solution column so that problems can be posted or exchanged for sharing.

| FACTS | SAMPLE QUESTIONS | SOLUTION |
|---|---|---|
| The Boston Marathon had 6674 entrants in '83, 6924 in '84, and 5595 is '85. | What is the average number of entrants over the three year period? (nearest whole number) | 6674<br>6924<br>+ 5595<br>19193<br><br>$3\overline{)19193}$ = $6397\,{}^{2}/_{3}$<br>6398 Entrants |

<u>FURTHER DISCUSSION / FOLLOW-UP:</u> Put some constraints on the kinds of questions that can be used. For example:

Must involve more than one operation for solution.
Must involve a particular operation (s).
Must involve a percent or decimal.
Must have extraneous data among facts.

This can be used as a bulletin board activity where the teacher posts one or more facts on a regular basis and students contribute questions (and solutions). Plan a fact or data gathering field trip to the Boston Marathon. The problems generated could be put on 3 X 5 cards (with answers on the back) to become a problem solving deck. See other <u>What's the Problem at</u> _____ activities.

The *Standards* state that in grades 5–8 the mathematics curriculum should include numerous and varied experiences with problem solving as a method of inquiry and application so that students can–

- Use problem-solving approaches to investigate and understand mathematical content;

- Formulate problems from situations within and outside mathematics;

- Develop and apply a variety of strategies to solve problems, with emphasis on multistep and non-routine problems;

- Generalize solutions and strategies to new problem situations;

- Acquire confidence in using mathematics meaningfully.

# Unified Science and Mathematics for Elementary Schools (USMES)

1.  LOG ON TRAFFIC FLOW

by John Flores*
Hosmer School, Grade 5
Watertown, Massachusetts
(September 1972-June 1973)

ABSTRACT

Children in this fifth-grade class concentrated on how they could improve traffic flow and pedestrian safety in a local center, Watertown Square. After drawing a map of the area and listing questions to be answered in their investigations, the students divided into groups to interview people about the traffic problem, to collect traffic data, and to photograph and film the area. The data group measured the widths of streets and traffic islands, timed traffic lights and pedestrians crossing at different points, and counted cars coming into the square. When the groups had gathered enough information, they plotted their results on graphs. Then they returned to the square to collect additional information that they thought was needed. The children discussed what changes they felt should be implemented, including overpasses, a tunnel, and an additional street through the central island. Each group made a model of the square incorporating their recommended changes. The models were displayed in the school, and the local paper published an article on the children's research and conclusions.

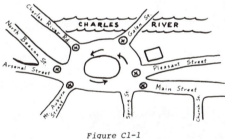

Figure C1-1

I introduced Traffic Flow to my class during a discussion of the problems they encountered riding their bicycles and traveling by car with their parents through Watertown Square. The children came up with the following list of problems:

1.  Traffic does not flow continuously.
2.  Parking, both legal and illegal, slows the flow of traffic.
3.  Pedestrians don't have enough time to cross.
4.  Lights are timed badly; some are too long, others too short.

## Traffic Flow

"Traffic Flow," like all of the USMES lessons, provides an engaging problem-solving activity in which students experience the vital role that mathematics plays in our society.

Students are asked to consider the safety and efficiency of several local traffic intersections and identify one location which they think could be improved. Once they have selected a site, they hold class discussions to determine the types of data that they will need to prepare a recommendation. Working in groups, they gather data via surveys, measurements and observations. They may also take photos or videos for documentation. Once the data has been gathered, they graph and analyze their results and prepare a report for local traffic officials in which they suggest ways to solve the problem and improve the flow of traffic.

- For further information about USMES turn to pages 3, 55 and 115.

From USMES, "Traffic Flow." USMES materials were developed by the Education Development Center. For information regarding USMES contact The Learning Team.

*Communicating mathematically requires students to reach agreement about the meaning of words and to recognize the importance of shared definitions. As students' ability in mathematics grows, so should their ability to communicate mathematically. The lessons below provide interesting opportunities for students to describe mathematically and/or interpret a variety of situations.*

# Mathematics Curriculum and Teaching Program (MCTP)

## Tell Me a Story

As its name implies, "Tell Me a Story" encourages students to discuss and then act out a story based on their interpretation of a graph. The graph represents cars travelling on a road.

The teacher begins the lesson by demonstrating techniques for reading and interpreting graphs. The students are asked to figure out what a graph representing a particular part of a road over time tells them about the movements of one car or several cars. Working in small groups they discuss and interpret the graph. They are asked to explain what has happened, either in writing or in a skit, and what a passenger in a car might have seen and done.

As an extension activity, students may prepare a computer simulation of their interpretation of the graph. Additional extensions include creating a second graph or extending the original.

• For further information about MCTP turn to pages 10, 16, 51, 74 and 102.

From MCTP, <u>Activity Bank,</u> Volume 1, Chapter 5, "Tell Me a Story," pages 253-256. MCTP materials were assembled by MCTP and are available from the Curriculum Corporation.

### 4. Tasks for each group

After ten minutes, allot a car to each group in the class.

> YOUR GROUP ARE ALL TRAVELLING IN THE BLUE CAR. WHAT CAN YOU SEE GOING ON AROUND YOU ... IN FRONT OF THE CAR? ...THROUGH THE REAR VISION MIRRORS?

> PREPARE A GROUP REPORT GIVING THE HIGHLIGHTS OF YOUR TRIP FOR THESE FOURTEEN SECONDS. YOU HAVE TEN MINUTES TO PREPARE THE REPORT. THEN EACH GROUP WILL HAVE ABOUT TWO MINUTES TO PRESENT IT

> MAKE THEM INTERESTING, BUT ACCURATE

A sample report could be simulated by the teacher. Such a report could highlight possibilities but should not stifle creativity. Make sure there is enough time for the reports — ten or fifteen minutes for reporting should suffice.

### 5. Group reports

From your visit to each group, you can make a judicious decision on which group to invite to present the first report.

There is probably no need for every group to present a report. Some will be enthusiastic to do so, but after three reports, remaining groups may prefer to supply additional observations.

> FOR THE REST OF THE CLASS, YOUR TASKS ARE TO LISTEN FOR ANY INACCURACIES — ESPECIALLY FOR ANY REFERENCES TO THE PARTICULAR CAR TO WHICH YOU WERE ASSIGNED, AND TO JUDGE THE OVERALL ACCURACY OF THE REPORTS

While pupils are preparing reports the teacher should visit each group to:
• ensure that ideas are being recorded,
• encourage a balance between creativity and mathematical accuracy,
• stimulate ideas if discussion is slowing down, and encourage all members of the group to share in the presentation, rather than leave it to one spokesperson.

The assessment used here is informal and non-threatening as the teacher encourages tactful appraisal of the strengths (and any weaknesses of each report).

Non-competitive assessment allows many girls to feel more comfortable, particularly when considering the subject matter.

### Sample reports

**a.** Pupils presented the report with their chairs arranged to simulate car seats. Each pupil contributed comments.

*'We left the telephone box and jumped into our green Ferrari and raced off towards home at 36 metres per second (108 km/h).*
*That blue car twenty metres ahead of us is going pretty slowly!'*
*Time = 1. (intoned by the car microcomputer)*
*'As we passed the blue car (30 metres from the phone box) we could see an orange car up ahead, a red car parked by the side of the road, and 50 metres ahead there was this large black truck coming towards us.'*
*Time = 2.*
*'Did you see that orange car nearly sideswipe the parked car as it passed?'*
*Time = 3 and a bit.*
*'Hey, look! The red car belongs to Mr. Smith our maths teacher. He's got a flat tyre' (sympathetic mutterings) ... (The saga continues.)*

**b.** Applying a different approach, one group used the chalkboard ledge and coloured pens to produce a 'puppet show' of the cars in motion.

### Extensions
• One school reported some keen pupils producing a computer-graphics picture of the cars in motion.
• Interested groups may like to create another graph, or extend the worksheet graph.

**The *Standards* state that in grades 5–8 the study of mathematics should include opportunities to communicate so that students can–**

- Model situations using oral, written, concrete, pictorial, graphical and algebraic methods;

- Reflect on and clarify their own thinking about mathematical ideas and situations;

- Develop common understanding of mathematical ideas, including the role of definitions;

- Discuss mathematical ideas and make conjectures and convincing arguments;

- Appreciate the value of mathematical notation and its role in the development of mathematical ideas;

- Use the skills of reading, listening, and viewing to interpret and evaluate mathematical ideas.

# Used Numbers: Real Data in the Classroom

## HOW MUCH TALLER IS A FOURTH (FIFTH, SIXTH) GRADER THAN A FIRST GRADER?

### INVESTIGATION OVERVIEW

#### What happens

Students measure their own height and the height of students in a first grade class, compare the heights of the two classes, and find ways to describe and represent the comparison. With this problem, you and your students move into a more complex investigation requiring several class sessions.

The activities take four class sessions of about 40 minutes each (minimum). One way of breaking the investigation into four sessions is suggested in the following outline; you can vary this as appropriate for your students. If students decide to remeasure the class or to take measurements of more classes, you may need an additional session. Also, some students may need more time to finish their presentation graphs.

#### What to plan ahead of time

▼ Provide measuring tools—yardsticks, metersticks, or tape measures—for each group of students (Sessions 1 and 3).

▼ Have calculators available.

▼ Provide unlined paper for sketching the data.

▼ Duplicate Student Sheet 1 (page 69) for each small group (Session 2).

▼ Arrange with the first grade teacher(s) a way for your students to collect the heights of first graders (Session 3).

▼ Provide materials for making presentation graphs: squared paper with inch squares or centimeter squares (a reproducible sheet of centimeter squares is

provided on page 79), and colored markers or crayons (Session 4).

▼ Copies of the height data (for both their own class and the first graders) for each small group (Sessions 2, 3, and 4).

#### Important mathematical ideas

**Inventing ways to quantify differences between two sets of data.** In order to compare two data sets, students use the ideas they have developed about describing the shape of the data and summarizing what is typical of a particular set of data. Comparison motivates students to capture the data in a single number or a small interval ("Fourth graders are about 55 inches tall" or "Fourth graders are between 54 and 56 inches tall") without ignoring the context for that number

## How Much Taller is a Fourth (Fifth, Sixth) Grader than a First Grader?

In this lesson, students practice a variety of communication skills. As a group, they discuss measurement techniques, brainstorm ways to describe data, formulate problems, and decide on problem-solving strategies.

Students begin the lesson by collecting data. They measure their own heights and the heights of students in a first-grade class. Once the information is collected, they brainstorm about the variety of ways to organize and describe the data, including line graphs and tallies. They then interpret the data, formulating and solving problems by comparing heights in the two classes.

- For further information about Used Numbers turn to pages 7 and 119.

From Used Numbers, Statistics: The Shape of the Data, Part 2: "How Much Taller is a Fourth (Fifth, Sixth) Grader than a First Grader?" pages 35-42. The Used Numbers series, written by Technical Education Research Center (TERC), Lesley College, and the Consortium for Mathematics and Its Applications (COMAP), is available from Dale Seymour Publications.

*Conjecturing and demonstrating the validity of conjectures is the essence of the creative act in mathematics. The ability to explore, conjecture, and validate is critical to the development of mathematical reasoning. The lessons below illustrate how reasoning can be an integral part of mathematical activity.*

## Mathematics Resource Project (MRP)

### Reflection Methods

"Reflection Methods" helps students understand and apply reasoning processes, with special attention to spatial reasoning and reasoning with proportions.

In this lesson, which is taken from a larger unit entitled "Geometry and Visualization," students develop their visual reasoning skills. In a series of exercises they learn to recognize and draw reflections using a number of different materials, including grid paper, tracing paper, folded paper and plastic which they place along the line of reflection perpendicular to the paper. They also learn to approximate the line of reflection within a pair of figures as well as about translations or slide motions involving both horizontal and vertical shifts.

• For further information about MRP turn to pages 52 and 103.

From the Mathematics Resource Project, Geometry and Visualization, "Reflection Methods," pages 310-317. Mathematics Resource Project materials were developed at the University of Oregon and are available from Creative Publications.

## REFLECTION METHODS

A) A method for drawing reflections uses grid paper. See *Grid Reflections*.

B) Two other methods use tracing paper.

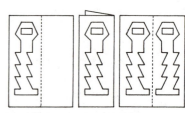

   1) Have students trace the figure and the line of reflection on tracing paper. Then fold the paper along the line of reflection with the figure on the outside. Tracing over the figure produces the reflection which can be seen by unfolding the paper.

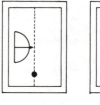

   2) Mark a reference dot on the line of reflection. Place tracing paper over the figure and trace it, the line of reflection, and the reference dot. Flip the tracing paper across the line of reflection and line up the lines and the reference dots. Retrace the figure pressing down hard to leave an impression on the original paper. Remove the tracing paper and mark over the impression.

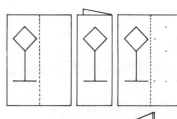

C) Fold the paper along the line of reflection. Then using a pin or compass point punch tiny holes through the vertices of the figure. Unfold the paper and connect the pin holes to produce the reflection. More pin holes will be necessary if the figure contains curved lines.

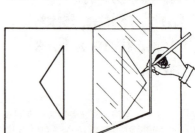

D) Place a piece of plastic along the line of reflection perpendicular to the paper. Look through the plastic and with a pencil carefully trace the image as it appears through the plastic. Straight line figures can be done by marking the vertices and then connecting the points.

The *Standards* state that in grades 5–8 reasoning shall permeate the mathematics curriculum so that students can–

- Recognize and apply deductive and inductive reasoning;

- Understand and apply reasoning processes with special attention to spatial reasoning and reasoning with proportions and graphs;

- Make and evaluate mathematical conjectures and arguments;

- Validate their own thinking;

- Appreciate the pervasive use and power of reasoning as a part of mathematics.

# Comprehensive School Mathematics Program (CSMP)

```
ACTIVITY H11:    HAND-CALCULATOR PROBLEMS INVOLVING TWO OPERATIONS #1


PREREQUISITE:    Activity H8.


OBJECTIVE:       Students will practice skills in pattern recognition and in
                 making and testing hypotheses.


Draw this picture on the board.
```

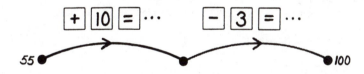

```
T:  What number could be here (point to the unlabeled dot)?  There are many
    possibilities.  Write at least one on your paper.  Try to find a pattern.


Let the class work on this problem for a few minutes.  Then begin to construct
a list of students' solutions on the board.  Perhaps 115 will be suggested.


T:  To get 115, how many times did you press [=] using the red arrow?  (Six)
    And how many times using the blue arrow?  (Five)


As the information is reported, trace the arrows and compute the middle number
to check the accuracy of the response.  Continue to add numbers to your list
as correct suggestions are verified.  Note that incorrect suggestions should
be checked to verify their incorrectness, just as correct answers are
verified.  Continue until sufficient numbers are on the list to investigate
possible patterns.  At this point accept any reasonable explanation, because
the idea may be difficult to verbalize.  If no explanation is forthcoming, do
not be concerned.  Do not force an explanation at this early stage.  A
discussion of the +30 pattern similar to the following is possible.
```

## Hand-Calculator Problems Involving Two Operations

"Hand-Calculator Problems Involving Two Operations" helps students make and evaluate mathematical conjectures and arguments.

In this lesson, students practice recognizing patterns and making and testing hypotheses. The generating problem challenges them to use their calculators to describe all possible ways to go from 55 to 100 through combinations of successive additions by 10 and subtractions by 3. They build a table representing the different ways and then describe the pattern. After that, they hypothesize about the comparable patterns that would arise if the 10 and 3 were changed to two other numbers.

Implicit in this lesson are the concepts of variables and equations with two unknowns.

- For further information about CSMP turn to pages 11, 23, 26, 38 and 91.

From CSMP, <u>Activities for TOPS,</u> Strand III, Activity H11, pages 22-23. CSMP materials were created by, and are available from, McREL Publications.

*Students' intellectual perspective must include recognizing the connection between various mathematical ideas as well as between mathematics and other topics. These lessons help students understand how mathematical symbols and processes are used to describe and model real-world phenomena and communicate complex thoughts in a concise and precise manner.*

# Regional Math Network (RMN)

## Living on the Colony

"Living on the Colony" graphically illustrates the importance of mathematics in our society today and in the future.

In this module, which is part of the Math/ Space Mission unit, students design their own space colony. The lesson begins with a class discussion of the colony with particular attention to its energy needs.

Students begin by recognizing that the amount of energy needed by the colony must be estimated so that an appropriate number of solar panels can be provided. They project the number of watts used per month by a family of four. They then calculate the approximate number of watts used per day by a family, and project the needs of the entire colony. Finally, they calculate the dimensions for a variety of solar panels based on an earlier calculation.

- For further information about RMN turn to pages 30 and 110.

From Regional Math Network, Math/ Space Mission, Module IV, "Living on the Colony," pages 136-139. Regional Math Network materials were created at the Harvard Graduate School of Education and are distributed by Dale Seymour Publications.

---

### LIVING ON THE COLONY

| Preparation/Materials | Math Skills |
|---|---|
| • Solar Cells | • Estimation in Computation Measurement<br>• Area |

During class:
- Discuss the features that must be planned for living on a colony:
  - Is there gravity? How will it be produced?
  - Is there air? How will it be created?
  - What about energy?
  - What percent of space should be allocated for various functions:
    - living
    - exercise and recreation
    - agriculture
    - technical laboratories

- Focus on the energy issue. Discuss that energy will be obtained from the sun, using solar panels.
  The panels collect and transform solar energy to electrical energy.

- Point out to the students that the amount of energy needed by the colony must be estimated, so that an appropriate number of solar panels can be provided.

- Explain that the students must estimate the number of solar panels needed to meet the energy needs of the colony on a daily basis.
  - First, students must project the number of watts used per month for a family of 4.
  - Next, students calculate the approximate number of watts used per day by a family.
  - Finally, students should project the needs for the entire colony.

### Extensions
- Solar Cells
- Explore the other features of living on the colony: gravity, "air," use of space.

The *Standards* state that in grades 5–8 the mathematics curriculum should include the investigation of mathematical connections so that students can–

- See mathematics as an integrated whole;

- Explore problems and describe results using graphical, numerical, physical, algebraic and verbal mathematical models or representations;

- Use a mathematical idea to further their understanding of other mathematical ideas;

- Apply mathematical thinking and modeling to solve problems that arise in other disciplines such as art, music, psychology, science and business;

- Value the role of mathematics in our culture and society.

# The Language of Functions and Graphs

## A1    INTERPRETING POINTS

As you work through this booklet, discuss your answers with your neighbours and try to come to some agreement.

### 1. The Bus Stop Queue

Who is represented by each point on the scattergraph, below?

Alice    Brenda    Cathy    Dennis    Errol    Freda Gavin

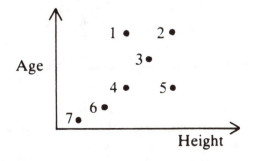

## Interpreting Points

"Interpreting Points" develops students' ability to reason from graphs and validate their own thinking.

Five situations from everyday life are used to explore how even a quick-sketch graph can be used to communicate a great deal of information. Students work in pairs and decide what the various graphs are communicating. They explain their reasoning and receive feedback from others.

The situations involve progressively sophisticated ideas, from straight forward comparisons of positions to comparisons involving gradients and eventually to the consideration of correlation and functional relationships. Detailed teaching suggestions and solutions are provided.

- For further information about The Language of Functions and Graphs turn to pages 44, 60, 67, 69 and 100.

From Shell, The Language of Functions and Graphs, Unit A1: "Interpreting Points," pages 64-73 and 100-101. The Language of Functions and Graphs materials were developed by, and are available from, the Shell Centre for Mathematical Education at the University of Nottingham in the United Kingdom.

*Building a sense of numbers and a facility with multiple forms of representation such as fractions, ratios, decimals and percents is a fundamental step in the learning process. The lessons below help students gain flexibility as well as grasp the appropriateness of each representation.*

# Comprehensive School Mathematics Program (CSMP)

## Percent #1

In "Percent #1," students practice representing and using numbers in three equivalent forms – integers, fractions and percents.

Percents are introduced in the context of exploring the composition of functions represented pictorially with arrow diagrams. For example, students look at how they might get from 48 to 36 by multiplying by a number, then dividing by another. They build an arrow diagram and table representing all of the different ways to do this. Percent is introduced from cases where the division part of the composition is divisible by 100. Students use patterns to find several percent calculations, then determine how much a 15% tip would be for each of several restaurant meals.

• For further information about CSMP turn to pages 11, 23, 26, 35 and 91.

From CSMP, Mathematics for Intermediate Grades, Part IV, Activity N7: "Percent #1," pages 41-48. CSMP materials were created by, and are available from, McREL Publications.

Draw arrows between some pairs of equivalent fractions to illustrate a technique for checking equality. For example,

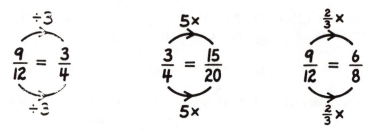

Discuss how to check that other of the fractions listed are equivalent.

Present each of the following problems, asking the students to provide the missing numerator or denominator. Answers are in the boxes.

$$\frac{3}{4} = \frac{\boxed{30}}{40} \qquad \frac{3}{4} = \frac{\boxed{75}}{100}$$

$$\frac{3}{4} = \frac{300}{\boxed{400}} \qquad \frac{3}{4} = \frac{1}{\boxed{\frac{4}{3}}}$$

Return to the table and the arrow picture to include ÷100 followed by 75×, if it is not already listed. Erase all of the other arrows leaving this picture.

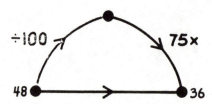

T:    We already know many names for the green arrow. These arrows suggest that the green arrow could be for $\frac{75}{100}$×. Another name for the green arrow is "75%" (read: seventy-five percent). "75%" is a name for ÷100 followed by 75× or 75× followed by ÷100.

The *Standards* state that in grades 5–8 the mathematics curriculum should include the continued development of number and number relationships so that students can–

• Understand, represent and use numbers in a variety of equivalent forms (integer, fraction, decimal, percent, exponential and scientific notation) in real-world and mathematical problem situations;

• Develop number sense for whole numbers, fractions, decimals, integers and rational numbers;

• Understand and apply ratios, proportions and percentages in a wide variety of situations;

• Investigate relationships among fractions, decimals and percents;

• Represent numerical relationships in one- and two- dimensional graphs.

# Middle Grades Mathematics Project (MGMP)

| TEACHER ACTION | TEACHER TALK | EXPECTED RESPONSE |
|---|---|---|
| On the board draw a picture of Pac Man with roughly a 10–15 cm diameter. | We are beginning a unit on similarity. | |
| | What do you think we mean when we use the word *similar*? | Two things are the same, or almost the same. |
| | Give me an example. | People are similar, cars, houses, clothes. |
| | Today we are going to learn about similarity using an ingenious little machine. | |
| Hold up two identical rubber bands. Knot two ends together. | It has two parts: We will knot the rubber bands like this. | |
| | This machine is called a *variable tension proportional divider*. We will just call it a *stretcher*. | |
| Point to Pac Man. | What is this strange figure? | Pac Man. Most students will recognize the figure from playing video games. |
| Demonstrate the machine. | Put one end of rubber band on the paper to the left of the figure far enough away so that the knot is on the figure. Call this point the *anchor point*. | |
| | Put pencil in the other end. Move pencil letting the knot trace the figure. Don't slack up on the first rubber band. | |
| | What happens? | We get another Pac Man similar to the original Pac Man, but bigger. |
| | What do we call this? | An enlargement. |

## The Variable Tension Proportional Divider

This lesson gives students an opportunity to work with the concepts of ratio, proportion, and percent.

Students make a variable tension proportional divider (stretcher) out of two rubber bands and a pencil. They use this stretcher to draw a new figure that is similar to the original, with corresponding

lengths that are twice as long and whose area is four times as large. They use the stretcher to enlarge a variety of shapes; a Pac Man, an apple, an old-fashioned car, and a football helmet.

The term "similar" has a precise meaning and the activities in this unit give students a chance to discover the mathematical meaning of similarity.

• For further information about MGMP turn to pages 41 and 104.

From MGMP, Similarity and Equivalent Fractions, Activity 1 : "The Variable Tension Proportional Divider," pages 5-13. MGMP materials were written at the Department of Mathematics of Michigan State University and are distributed by Addison-Wesley Publishing Company.

*It is important for students to understand that the underlying structure of mathematics is made up of many facets which blend together to make an integrated whole. The lessons below illustrate this cohesiveness and help students develop a basic understanding of number theory concepts.*

## Developing Mathematical Processes (DMP)

### Decimal Fractions

This lesson helps students develop and use order relations for whole numbers, fractions, decimals, integers and rational numbers.

"Decimal Fractions" is part of a larger unit on decimal fractions. The students write decimals to represent pictures involving thousandths, as shown here. In the variety of tasks included in this activity they are asked to shade the areas that a given decimal may represent, name places and digits, write decimals from a grid, and order decimals. Later in the activity, they use numbers that have digits only to the tenths place and only to the hundredths place.

• For further information about DMP turn to pages 19, 20, 43, 45 and 94.

From DMP, Book 74, Activity 74F: "Decimal Fractions," pages 24-26. DMP was developed at the University of Wisconsin-Madison and is published by Delta Education.

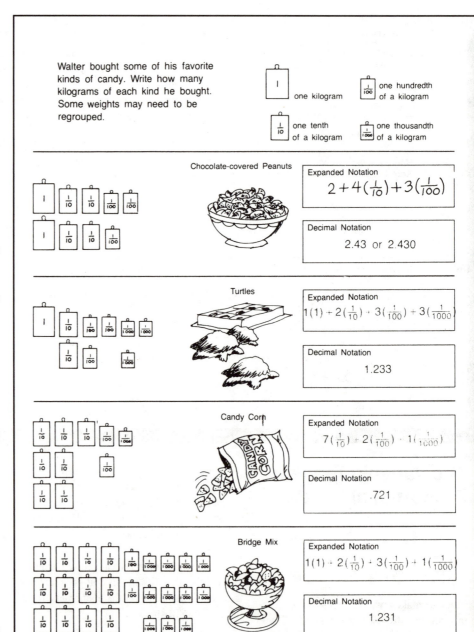

Walter bought some of his favorite kinds of candy. Write how many kilograms of each kind he bought. Some weights may need to be regrouped.

one kilogram

one tenth of a kilogram

one hundredth of a kilogram

one thousandth of a kilogram

**Chocolate-covered Peanuts**

Expanded Notation
$$2 + 4\left(\tfrac{1}{10}\right) + 3\left(\tfrac{1}{100}\right)$$

Decimal Notation
2.43 or 2.430

**Turtles**

Expanded Notation
$$1(1) + 2\left(\tfrac{1}{10}\right) + 3\left(\tfrac{1}{100}\right) + 3\left(\tfrac{1}{1000}\right)$$

Decimal Notation
1.233

**Candy Corn**

Expanded Notation
$$7\left(\tfrac{1}{10}\right) + 2\left(\tfrac{1}{100}\right) + 1\left(\tfrac{1}{1000}\right)$$

Decimal Notation
.721

**Bridge Mix**

Expanded Notation
$$1(1) + 2\left(\tfrac{1}{10}\right) + 3\left(\tfrac{1}{100}\right) + 1\left(\tfrac{1}{1000}\right)$$

Decimal Notation
1.231

The *Standards* state that in grades 5–8 the mathematics curriculum should include the study of number systems and number theory so that students can–

- Understand and appreciate the need for numbers beyond whole numbers;

- Develop and use order relations for whole numbers, fractions, decimals, integers and rational numbers;

- Extend their understanding of whole number operations to fractions, decimals, integers and rational numbers;

- Develop and apply number theory concepts (e.g., primes, factors and multiples) in real-world and mathematical problem situations.

# Middle Grades Mathematics Project (MGMP)

| TEACHER ACTION | TEACHER TALK | EXPECTED RESPONSE |
|---|---|---|
| | Clear your calculator. What will happen if we press 0 + 5 $=$ $=$ $=$ and so on? | We'll get 5, 10, 15, 20, and so on. |
| | Try it. | |
| | What did you get? Tell me how they are multiples. | $5 \times 1 = 5, 5 \times 2 = 10, 5 \times 3 = 15$, and so on. |
| | What would you do to get 3, 6, 9, 12, 15, and so on? | Press 0 + 3 $=$ $=$ $=$ and so on. |
| Display a transparency of Worksheet 7-1, 100 Board, on the overhead. Give the class instructions, in appropriate colors, on the overhead. The overall scheme for marking the 100-Board is illustrated in this square. With older students the marks alone will suffice without coloring in the triangular areas. Whichever scheme you use, have students record in the square provided on Worksheet 7-1 the code used to make the sieve. The follow-up questions make it *necessary* for students to be able to read the code on a particular number after the sieve is finished. | Cross out the 1. One is not a prime. The next number is 2. It is a prime. Circle it in red. What are the multiples of 2? Shade the upper left corner of the other multiples of 2 in red. Do not shade 2 itself. | 2, 4, 6, 8, and so on. and so on. |
| | Where are the multiples of 2? | They're in columns. |
| | We'll say that the shaded numbers fell through our sieve. What numbers are they? | The even numbers (except for 2). |
| | The next number is 3. What are the multiples of 3? Circle 3 in green. | 3, 6, 9, 12, and so on. |
| | Starting with 6, shade the other multiples of 3 using your green pen; shade the upper right corner. | and so on. |
| | Where are the multiples of 3? | On diagonals. |
| | The next number is 4. It's already been shaded. Why? | It is a multiple of 2. |

# Sifting for Primes

"Sifting for Primes" helps students understand and apply number theory concepts to real-world and mathematical problem situations.

Students explore the classic "Sieve of Eratosthenes" (shown above) to obtain prime numbers. When created with colored pens, however, the sieve also contains much information about factors, multiples, and composite numbers.

The goals of the unit are to define prime and composite numbers, to review the relationship between factors and multiples, to use the technique of sifting to find primes, and to use square numbers to predict the last number that must be sifted to find all primes less than a given number.

- For further information about MGMP turn to pages 39 and 104.

From MGMP, Factors and Multiples, Activity 7: "Sifting for Primes," pages 87-101. MGMP materials were written at the Department of Mathematics, Michigan State University, and are distributed by Addison-Wesley Publishing Company.

*Students need to develop a variety of computation and estimation skills including mental calculation, paper-and-pencil computation and the use of calculators and computers. These lessons help students understand several computation and estimation techniques.*

# University of Illinois Arithmetic Project (UIAP)

## Ways to Find How Many

"Ways to Find How Many" helps students develop, analyze, and explain procedures for computation and techniques for estimation.

The introductory activities in this lesson familiarize students with the length of a centimeter and the size of a centimeter cube, as shown on this page. Students are then asked to find the number of cubes contained in several different patterns, first by counting and then by using multiplication and other techniques. They practice making estimates of the number of cubes in a pattern and check their estimates. Next, they estimate and find the areas of a variety of geometric figures. Finally, they compute the number of centimeter cubes contained in stacks and towers of cubes.

• For further information about UIAP turn to pages 27 and 117.

From UIAP, Lesson 1: "Ways to Find How Many." The UIAP materials were developed at the Education Development Center, Inc. For further information contact The Learning Team.

## WAYS TO FIND HOW MANY

by David A. Page

This curve is four centimeters long:

So is this one:

Use any method you can think of to decide whether each of the following curves is shorter than four centimeters or longer than four centimeters. Don't bother to tell how many centimeters long they are. Just say *"longer* than four centimeters" or *"shorter* than four centimeters."

4. _____ than four centimeters

5. _____ than four centimeters

6. _____ than four centimeters

7. _____ than four centimeters

8. _____ than four centimeters

The *Standards* state that in grades 5–8 the mathematics curriculum should develop the concepts underlying computation and estimation in various contexts so that students can–

• Compute with whole numbers, fractions, decimals, integers and rational numbers;

• Develop, analyze, and explain procedures for computation and techniques for estimation;

• Develop, analyze and explain methods for solving proportions;

• Select and use an appropriate method for computing from among mental arithmetic, paper-and pencil, calculator and computer methods;

• Use computation, estimation and proportions to solve problems;

• Use estimation to check the reasonableness of results.

# Developing Mathematical Processes (DMP)

## Ratios and Proportions

"Ratios and Proportions" gives students a number of diverse opportunities to identify, analyze and explain methods for solving proportions.

In this lesson the term proportion is introduced. Students work with proportions in a variety of activities such as increasing recipes proportionately as shown here. These situations in which proportions are used are not new to students but simply another way to look at and solve a problem. Throughout this activity, students are asked to identify, read, write, and solve proportion sentences.

• For further information about DMP turn to pages 19, 20, 40, 45 and 94.

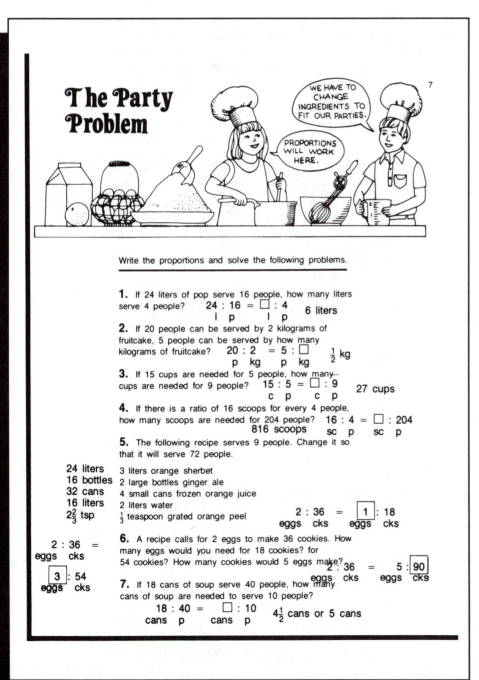

From DMP, Book 88, Activity 88C: "Ratios and Proportions," pages 14-17. DMP was developed at the University of Wisconsin-Madison and is published by Delta Education.

*The ability to describe, analyze and extend patterns and functional relationships is one of the most important and basic concepts that students need to master as they develop mathematical literacy. In the lessons below, students are asked to explore numerical patterning and to focus on building mathematical models that predict the behavior of real-world phenomena.*

## The Language of Functions and Graphs

### Sketching Graphs from Words

In "Sketching Graphs from Words" students create graphs, look for patterns, and analyze functional relationships to explain how a change in one quantity results in a change in another.

Several real-world situations are provided and students are asked to translate verbal descriptions into graphs. Two kinds of verbal forms are used: "full descriptions," which give an explicit account of how the variables relate to each other; and "trigger phrases," which ask the students to imagine a situation and then decide for themselves the nature of the relationship between the variables.

Students work in small groups discussing and sketching each situation. Once each group has arrived at a consensus, the entire class comes together for a discussion. Detailed teacher's notes and solutions are provided.

• For further information about The Language of Functions and Graphs turn to pages 37, 60, 67, 69 and 100.

From Shell, <u>The Language of Functions and Graphs</u>, Unit A3:"Sketching Graphs From Words," pages 82-87 and 102. <u>The Language of Functions and Graphs</u> materials were developed by, and are available from, the Shell Centre for Mathematical Education at the University of Nottingham in the United Kingdom.

### A3 SKETCHING GRAPHS FROM WORDS

**Picking Strawberries**

The more people we get to help, the sooner we'll finish picking these strawberries.

STRAWBERRY PICKERS WANTED

* **Using axes like the ones below, sketch a graph to illustrate this situation.**

Total time it will take to finish the job

Number of people picking strawberries

* **Compare your graph with those drawn by your neighbours. Try to come to some agreement over a correct version.**

* **Write down an explanation of how you arrived at your answer. In particular, answer the following three questions.**

   — **should the graph 'slope upwards' or 'slope downwards'? Why?**

   — **should the graph be a straight line? Why?**

   — **should the graph meet the axes? If so, where? If not, why not?**

The *Standards* state than in grades 5–8 the mathematics curriculum should include explorations of patterns and functions so that students can–

- Describe, extend, analyze and create a wide variety of patterns;

- Describe and represent relationships with tables, graphs and rules;

- Analyze functional relationships to explain how a change in one quantity results in a change in another;

- Use patterns and functions to represent and solve problems.

# Developing Mathematical Processes (DMP)

## Number Patterns from Geometry

Have you ever noticed that the ten pins in bowling are placed to form a triangle? Do you see that taking any number of rows also gives triangles?

The ten bowling pins are in four rows. To make the fifth row, you would need to add five more pins, but you would still get a triangle.

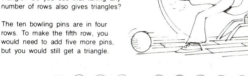

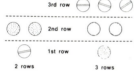

2 rows     3 rows

5 rows

**1.** How many pins are there altogether in the five-row triangle? **15**

**2.** How many pins would there be altogether in a six-row triangle? **21**

The total number of pins for each triangle is called a **triangular number**. Triangular numbers form a pattern, especially when you think of adding up the number of pins in the rows, one row at a time.

**3.** The pattern for triangular numbers in the chart at the right tells you that the tenth triangular number is
$1 + 2 + 3 + 4 + 5 + 6 + 7 + 8 + 9 + 10$
What is the tenth triangular number? **55**

| Sum of pins in the rows | Triangular number |
|---|---|
| 1 | 1 |
| 1 · 2 | 3 |
| 1 · 2 · 3 | 6 |
| 1 · 2 · 3 · 4 | 10 |
| 1 · 2 · 3 · 4 · 5 | 15 |
| 1 · 2 · 3 · 4 · 5 · 6 | 21 |

The total number of pins needed to form squares also makes a pattern. Look at the number of pins you must add each time to get the next larger square.

2 · 2     3 · 3     4 · 4
4 pins in all    9 pins in all    16 pins in all

**1.** Draw a 5 · 5 square on your paper. How many more pins does it have than the 4 · 4 square? **9**

**2.** How many pins in all are needed for a 7 · 7 square? **49** for a 9 · 9 square? **81**

The total number of pins for each square are called **square numbers**. The pattern for square numbers shows up nicely when you write them as a sum of the pins used as you go from one square to the next.

| Sum of pins in the rows | Square number | |
|---|---|---|
| 1 | 1 | 1 · 1 |
| 1 · 3 | 4 | 2 · 2 |
| 1 · 3 · 5 | 9 | 3 · 3 |
| 1 · 3 · 5 · 7 | 16 | 4 · 4 |
| 1 · 3 · 5 · 7 · 9 | 25 | 5 · 5 |

**3.** The eighth square number is the sum of the first eight odd numbers:
$1 + 3 + 5 + 7 + 9 + 11 + 13 + 15.$

Check to see that this sum gives you the same answer as 8 × 8. Now add up the first twelve odd numbers to see if that sum is 12 × 12. Add up the first fifteen odd numbers. Is that sum 15 × 15? **yes**

**4.** Use the addition pattern and multiplication to find the twentieth square number. **400**

**5.** The sum of the first five counting numbers gives you the fifth triangular number. The sum of the first five odd counting numbers gives you the fifth square number. Can you make a geometric figure from the number of pins that is the sum of the first five even counting numbers?

Figures will vary.

30 pins

## Patterns

"Patterns" helps students describe and represent relationships with tables, graphs and rules.

In this activity, which is part of a larger unit on patterns, students investigate number patterns that are either arithmetic or geometric. The patterns involve not only the numbers themselves but also the order in which they occur.

This activity helps students understand how the result in one quantity results in a change in another.

- For further information about DMP turn to pages 19, 20, 40, 43 and 94.

From DMP, Book 90, Activity 90F: "Patterns," pages 26-30. DMP was developed at the University of Wisconsin-Madison and is published by Delta Education.

*The concepts of variables, expressions and equations must be developed with care as students make the essential transition from arithmetic to algebra. In order to establish formal equation-solving procedures with a solid conceptual footing, it is important for students to explore them first in informal ways. The lessons below help students examine these algebraic concepts.*

# Madison Project

## Open Sentences

This lesson helps students understand the concepts of variables, expressions and equations. It helps them develop confidence in solving equations by using concrete, informal and formal methods.

"Open Sentences" introduces ordered pairs as solutions to open sentences. Students use tables, graphs and set notations to describe truth sets of equations for two variables. This lesson presents situations and number patterns with tables, graphs, and equations in order to explore the interrelationships of these representations.

• For further information about the Madison Project turn to pages 5, 14, 73, 85 and 101.

From the Madison Project, <u>Discovery in Mathematics</u>, Chapter 10: "Open Sentences with Two Placeholders," pages 75-78. Materials were created by the Madison Project and are published by Cuisenaire Company of America.

### OPEN SENTENCES WITH TWO PLACEHOLDERS*

The point of this chapter (and of the one following) is, of course, the traditional old notion of Cartesian co-ordinates used to display the graph of a function.

The modern language is much clearer than the traditional. If an open sentence has two (different) placeholders, its truth set will consist of *ordered pairs*.

Consider the open sentence $\triangle = \square + 1$. Since a triangle and a box are *different* shapes, we need not put the same number into both, although we *may* do so if we wish.

Each substitution must consist of a *pair* of numbers: one number to go into the box,

$$3 \to \square : \quad \triangle = \boxed{3} + 1$$

and one number to go into the triangle

$$7 \to \triangle : \quad \boxed{7} = \boxed{3} + 1.$$

Substituting the pair (3, 7) into the box and triangle, respectively, produces the *false* statement $7 = 3 + 1$, so we know that the pair (3, 7) does *not* belong to the truth set for the open sentence

$$\triangle = \square + 1.$$

Suppose we try the pair (5, 6). The traditional order is *first* to write the number that goes into the box, and *second* the number that is to go into the triangle. By writing the pair as (5, 6), we therefore mean:

$$5 \to \square$$
$$6 \to \triangle .$$

Substituting (5, 6) gives the *true* statement $\boxed{6} = \boxed{5} + 1$, so the pair (5, 6) *does* belong to the truth set of the open sentence

$$\triangle = \square + 1.$$

Notice that the *order* is important. The pair (5, 6) would mean (according to the order convention stated above):

$$5 \to \square$$
$$6 \to \triangle .$$

Hence it produces the *true* statement $\boxed{6} = \boxed{5} + 1$; consequently (5, 6) *does* belong to the truth set for $\triangle = \square + 1.$

**The *Standards* state that in grades 5–8 the mathematics curriculum should include explorations of algebraic concepts and processes so that students can–**

- Understand the concepts of variables, expressions and equations.

- Represent situations and number patterns with tables, graphs, verbal rules and equations and explore the interrelationships of these representations;

- Analyze tables and graphs to identify properties and relationships;

- Develop confidence in solving linear equations using concrete, informal and formal methods;

- Investigate inequalities and nonlinear equations informally;

- Apply algebraic methods to solve a variety of real-world and mathematical problems.

# Teaching Integrated Math and Science (TIMS)

# Rolling Along II

## Teacher Lab Discussion

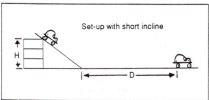

Set-up with short incline

Fig. 1

**Fig. 1** is a picture of the experiment that we call **Rolling Along II.** Its companion experiment, not surprisingly, is called **Rolling Along I.** The basic set-up is the same for both experiments. Books, wooden blocks, or Cube-O-Grams are used to sup-

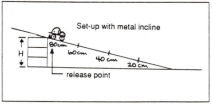

Set-up with metal incline

release point

Fig. 2

port an incline so that its back is a height H off the floor. The TIMS cart or a roller skate (the children can bring these in) is released from the top of the incline and rolls a distance D until it stops. In **Rolling Along I** the type of car is the manipulated variable and D is the responding variable. The value of H is held fixed. In **Rolling Along II,** H is the manipulated variable, D is the responding variable, and the type of cart is controlled. **Rolling Along I** is designed for 1st and 2nd graders while **Rolling Along II** is for 3rd grade and up.

For **Rolling Along II** (the only experiment we shall refer to in the rest of this write-up) the incline should be about 30cm long (this length is not cru-

cial, but we shall use an incline of this approximate dimension for several other experiments) and can be cut out of wood. Our teachers have found that local lumber yards are happy to cut wood for you, especially when you tell them it is for a science experiment! One can also use the 120cm metal incline that comes with the TIMS package of materials. as shown in **Fig. 2.** The <u>single</u> <u>release</u> point can be at 80cm, or further down at 60cm, or 40cm, depending on how much room the car has to roll.

If you are using the short ramp, then the values of 4cm, 8cm, and 12cm will do nicely. If you use the long ramp, use 8cm, 16cm, and 24cm for H in order to give the ramp a large enough slope so that we can minimize the effect of friction while the cart rolls down the incline. Using these values of H will leave the data points well-spaced for an accurate fit, as well as plenty of room for interpolation and extrapolation questions.

If your students are counting into the 100s, then you can tell them to measure the responding variable in cm. For our cart, D was about 250cm when H was 8cm. The children should take 3 "runs" at each value of H . Let each partner take turns releasing the cart and measuring D. For any given value

| H (cm) | D (cm) | | | |
|---|---|---|---|---|
| | Trial 1 | Trial 2 | Trial 3 | Average |
| 4 | 123 | 125 | 123 | 123 |
| 8 | 256 | 252 | 254 | 254 |
| 12 | 364 | 371 | 367 | 367 |

Fig. 3

of H , the measurement of D should be within 5cm to 10cm of the central value. The data we obtained using a short ramp is shown in **Fig. 3.**

## Rolling Along II

"Rolling Along II" helps students better understand the concepts of variables, expressions, and equations and builds their confidence in solving equations using concrete, informal and formal methods.

In this lesson students practice measuring lengths. They are asked to identify and recognize the manipulated variable and responding variable. Their measurements also supply the data necessary to incorporate the concepts of ratio/proportion, as well as a graphical representation of the information collected.

- For further information about TIMS turn to pages 2, 22, 54 and 114.

From TIMS, Experiment 203: "Rolling Along II." TIMS materials were created at, and are available from, the project office at the University of Illinois at Chicago.

*With the expanding use of modern technology, students must be comfortable with, and well versed in, the methods of collecting, organizing, representing and analyzing data. The lessons below help them develop an appreciation for statistical methods in the decision-making process.*

## Quantitative Literacy Series (QLS)

### The Ages of U.S. Presidents at Their Death

"The Ages of U.S. Presidents at Their Death" provides an opportunity for students to construct, read and interpret tables, charts and graphs.

Students construct a stem-and-leaf plot graph to quickly organize and display data about the ages of United States Presidents at their death. They analyze the data by observing the shape of the data set and by interpreting graph features such as extreme values, outliers, clusters, gaps and the relative positions of items. Students are asked to communicate their observations through written descriptions.

- For further information about QLS turn to pages 50, 76, 79 and 109.

From Quantitative Literacy Series, Exploring Data, Section II: "Stem-and-Leaf Plots: The Ages of U.S. Presidents at Their Death," page 13. QLS books were written by the Joint Committee on the Curriculum in Statistics and Probability of the American Statistical Association and the National Council of Teachers of Mathematics. QLS books are distributed by Dale Seymour Publications.

---

**Application 3**

---

**Ages of U.S. Presidents at Their Death**

The table below lists the presidents of the United States and the ages at which they died.

| | | | | | |
|---|---|---|---|---|---|
| Washington | 67 | Filmore | 74 | Roosevelt | 60 |
| Adams | 90 | Pierce | 64 | Taft | 72 |
| Jefferson | 83 | Buchanan | 77 | Wilson | 67 |
| Madison | 85 | Lincoln | 56 | Harding | 57 |
| Monroe | 73 | Johnson | 66 | Coolidge | 60 |
| Adams | 80 | Grant | 63 | Hoover | 90 |
| Jackson | 78 | Hayes | 70 | Roosevelt | 63 |
| Van Buren | 79 | Garfield | 49 | Truman | 88 |
| Harrison | 68 | Arthur | 57 | Eisenhower | 78 |
| Tyler | 71 | Cleveland | 71 | Kennedy | 46 |
| Polk | 53 | Harrison | 67 | Johnson | 64 |
| Taylor | 65 | McKinley | 58 | | |

1. Make a stem-and-leaf plot of the ages using these stems.

```
4 |
5 |
6 |
7 |
8 |
9 |
```

2. How many presidents died in their forties or fifties?

3. Who lived to be the oldest?

4. Label the four presidents who were assassinated.

5. What is the shape of this distribution?

6. Write a one-paragraph description of the information shown in the stem-and-leaf plot, including information about the presidents who were assassinated.

The *Standards* state that in grades 5–8 the mathematics curriculum should include exploration of statistics in real-world situations so that students can–

- Systematically collect, organize and describe data;

- Construct, read and interpret tables, charts and graphs;

- Make inferences and convincing arguments that are based on data analysis;

- Evaluate arguments that are based on data analysis;

- Develop an appreciation for statistical methods as powerful means for decision-making.

# Journeys in Mathematics

## Introduction

1. **Total the adult taste data and record the results in the sixth column of the class table.**

   Label this column "Taste Preferences of Adults."

   The total number of adult votes should be equal to the total number of student votes.

2. **Ask the class to compare the graphs on pages 14 and 33 of the student book.**

   These graphs are also shown on transparency 9A. Graph A is "Our Class' Favorite Pizza" from Activity 4 and Graph B is a new graph, "Favorite Pizza of Adults and Our Class."

   Students may note similarities and differences such as the following:

   - The types of pizza are the same.
   - The scale range is 0 to 9 on both graphs.
   - The scales are both marked in intervals of 1.
   - Graph B has two bars over each type of pizza.
   - Graph B has a *key* that tells us that the white bars are the adult votes and the black bars are the student votes. (In the student book, the bars representing the adult votes are purple, and the bars representing the student votes are blue.)
   - The left axis is labelled as the Number of Students on Graph A and as the Number of People on Graph B.

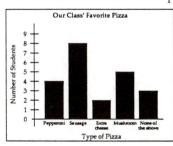

*Graph A (student book page 14)*

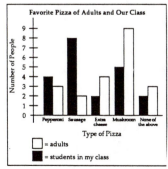

*Graph B (student book page 33)*

## Double Bar Graph

In "Double Bar Graph" students construct, read and interpret graphs. They learn to make inferences and arguments that are based on their data. Through this process, students develop an appreciation for statistical methods as a decision-making tool.

Students make a double bar graph using the data they collected from a survey conducted in a previous class. Column A is labeled "Our Class's Favorite Pizza," and Column B is labeled "Favorite Pizza of Adults and Our Class." Students then use the double bar graph to compare the two sets of data to determine the pizza preferences.

This lesson and its extensions help students practice conducting class surveys, organizing their results in tables, and representing and analyzing their data in double bar graphs.

- For further information about Journeys in Mathematics turn to pages 4, 24, and 97.

From Journeys in Mathematics, Data and Decisions, Section 2, Activity 9, pages 71-75. Developed by Education Development Center and published by WINGS for learning.

*To fully understand the relationship between the numerical expression of a probability and the situation that gave rise to those numbers, students need to explore probability models. The lessons below offer students opportunities to experience and appreciate the pervasive use of probabilities in today's world.*

# Quantitative Literacy Series (QLS)

## Knowing Our Chances in Advance

"Knowing Our Chances in Advance" is made up of six activities in which students explore the relationship between theoretical and estimated probabilities.

Working in small groups, students explore theoretical probabilities for equally likely outcomes. In the first activity, they spin a spinner with eight numbered sections 40 times, graph the results, and answer questions like: What is the probability that the spinner will land on number 3? Find the fraction of 3's observed in the total sample of 40 spins.

Later activities look at probabilities using a deck of cards, birth dates, experimenting with ESP (pictured here), playing chips, and two random number tables. By the end of the unit, students have defined the theoretical probability of a situation and explored its relationship to an estimated probability.

- For further information about QLS turn to pages 48, 76, 79 and 109.

From Quantitative Literacy Series, Exploring Probability: "Knowing Our Chances in Advance," pages 19-25. The QLS materials were developed by the Joint Committee on the Curriculum in Statistics and Probability of the American Statistical Association and the National Council of Teachers of Mathematics and are distributed by Dale Seymour Publications.

Do you have ESP (extrasensory perception)? Try this experiment and see.

A. Make a set of 40 cards of the same size using four different symbols, so that you have ten cards for each symbol. They might look like this:

| square | oval | plus | squiggle |

B. Choose a partner. Ask him or her to face away from you (or blindfold him or her).

C. Mix the cards well.

D. Turn over a card and concentrate on the symbol it shows.

E. Ask your partner to read your mind and tell you what is written on the card.

F. Record the answer *without* telling him or her whether or not it is correct.

G. Repeat the procedure and tally the results until you have recorded a total of 20 answers.

| Right Answer | Wrong Answer | Total |
| --- | --- | --- |
| | | |

1. Do you think your partner has ESP? Why or why not?

2. If your partner is just guessing, what is the probability of his or her guessing correctly on any one trial?

3. If you were to run this experiment again for 100 trials, about how many answers do you predict would be correct?

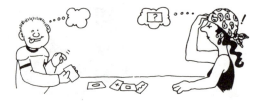

**The *Standards* state that in grades 5–8 the mathematics curriculum should include explorations of probability in real-world situations so that students can–**

- Model situations by devising and carrying out experiments or simulations to determine probabilities;

- Model situations by constructing a sample space to determine probabilities;

- Appreciate the power of using a probability model by comparing experimental results with mathematical expectations;

- Make predictions that are based on experimental or theoretical probabilities;

- Develop an appreciation for the pervasive use of probability in the real-world.

# Mathematics Curriculum and Teaching Program (MCTP)

## Mortality quiz

*When an insurance company charges one person $1.28 and another person $5.40 per month for the identical insurance policy you can bet your life\* that they have good reasons — mostly based on statistics. This activity looks at such figures and the reasons for them.*

*\* Couldn't resist the double entendre. Of course a life insurance policy is a bet with the insurance company on your life. (If you die, you [???] win the bet)*

> **WARNING**
> Since the subject is the statistics of death, be very sensitive to the feelings of children who may have had recent bereavements or whose family or friends have had serious injuries. If you know of such a case you may choose to reschedule this activity to a more appropriate time.

### 1. Safe ages and unsafe ages

The questions and the given commentary are designed to find out whether pupils are aware that different ages have different probabilities of dying.

Trials schools reported that this question need explanation.

The answer is based on the most recent figures available, males and females combined.

*I HAVE AN INTERESTING FIVE QUESTION QUIZ FOR YOU, BASED ON THE LIFE INSURANCE INDUSTRY. NOW YOU PROBABLY KNOW THAT THE INSURANCE COMPANIES COLLECT STATISTICS AND, INDEED, KNOW WHAT THE CHANCES ARE. WELL THEN, WHICH DO YOU THINK WOULD COST MORE TO INSURE — A 20 YEAR OLD PERSON OR A 95 YEAR OLD?*

*GOOD! NOW HERE IS QUESTION ONE. WRITE DOWN WHAT YOU THINK IS THE 'SAFEST AGE TO BE IN AUSTRALIA. WHAT DO I MEAN BY 'SAFEST AGE'? IT IS THE AGE AT WHICH THE CHANCES OF DEATH IN THE NEXT YEAR ARE THE LEAST. WHAT AGE DO YOU THINK IS SAFER THAN ALL OTHERS? NOW YOU KNOW THAT IT'S NOT 95, SO WHAT AGE IS SAFEST?*

*95 — BECAUSE THE 20 YEAR OLD PERSON SHOULD BE FIT*

### 2. The class perception of the safe age

Write on the chalkboard the spread and the frequency of answers. Then tell the class the answer.

Trials classes showed considerable variation, giving answers from one to fifty.

Guessing exposes current perceptions and allows them to be compared with reality.

This question is actually included in *Trivial Pursuit*.

*WHAT NUMBER DID YOU HAVE TRACEY?*

*37!*

*WHO HAD HIGHER?*

50 45 40 42

*WHO HAD LOWER?*

18 20 21 30

*WELL THE ANSWER IS TEN. WHEN YOU THINK ABOUT IT I'M SURE YOU CAN SEE THE LOGICAL REASON FOR THIS. WE'LL DISCUSS MORE OF THESE REASONS LATER*

## Mortality Quiz

"Mortality Quiz" offers students a chance to see how probability and mathematics can help them make choices in their lives.

In this lesson, students are first asked to guess the "safest age," that is, the age at which the chances of death in the next year are the least (shown here). Later in the lesson, they are given an unscaled graph of the probability of death at various ages. Students guess the ages that correspond to several critical points on the graph. They then discuss the social factors which contribute to the mortality rates at various ages, especially the late teens and early twenties. Extension activities include: comparing mortality rates between countries (the data in this lesson are for Australia), looking at trends over time, comparing different groups of people, and looking at insurance rates.

- For further information about MCTP turn to pages 10, 16, 32, 74 and 102.

From MCTP, Activity Bank, Volume I, Chapter 2, pages 125-131. Materials were assembled by MCTP and are published by the Curriculum Corporation.

# 5-8 Standard 12: Geometry

*The understanding of geometric figures and relationships is essential in interpreting many other mathematical concepts and real-world phenomena. The lessons below provide students with experience in geometric models and problem solving, helping to develop their spatial sense and facility in geometry.*

## Mathematics Resource Project (MRP)

### Symmetry and Motions

This group of activities on symmetry and motions is taken from a larger unit called "Geometry and Visualization." These activities provide students with opportunities to visualize and represent geometric figures, with special attention to developing spatial sense.

In this lesson students make and identify figures with line symmetry, learn how to determine the number of lines of symmetry, and discover the relationship between angles and lines of symmetry. They develop their spatial sense and visual reasoning skills as they recognize and draw reflections, approximate lines of reflection, and identify reflections across two perpendicular lines. Finally, students are introduced to slide motions through the use of grid paper.

- For further information about MRP turn to pages 34 and 103.

From Mathematics Resource Project, Geometry and Visualization, "Symmetry and Motion," pages 296-302. MRP materials were developed at the University of Oregon and are available from Creative Publications.

---

COMMENTARY                                                    SYMMETRY AND MOTIONS

#### MOTIONS

Three types of elementary motions of geometry are reflections (flips), rotations (turns) and translations (slides). Each of these has been applied to the ⌐F below.

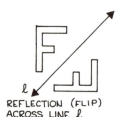

REFLECTION (FLIP)
ACROSS LINE ℓ.

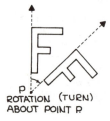

ROTATION (TURN)
ABOUT POINT P

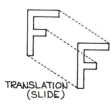

TRANSLATION
(SLIDE)

#### Reflection

Folding and cutting a figure with line symmetry is much like reflecting a shape across a line. (See *Reflection Methods*.) Each point P has a reflection point P' which is the same distance from the axis of reflection. This axis is the perpendicular bisector of line segment PP'. The student page *Reflections : Reflections* asks students to decide if one figure is a reflection of the other. The students could work the page informally by using a mirror, mira (a manipulative available from several commercial sources) or a piece of sheet plastic. More formally, they could use a ruler and a right angle to check for equal distance and perpendicularity. It is best to give students some experience with reflection instruments such as mirrors, miras or plastic pieces before asking them to use only visual means of recognizing reflections.

After students have learned to recognize the reflection of a shape, they can try various methods for drawing reflections. You might want to start students with grid paper reflections. Some fairly simple designs can be reflected <u>left and right</u> across a vertical line.

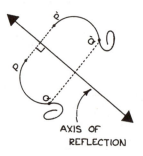

AXIS OF
REFLECTION

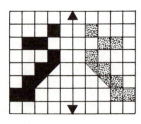

The *Standards* state that in grades 5–8 the mathematics curriculum should include the study of the geometry of one, two and three dimensions in a variety of situations so that students can–

- Identify, describe, compare and classify geometric figures;

- Visualize and represent geometric figures with special attention to developing spatial sense;

- Explore transformations of geometric figures;

- Represent and solve problems using geometric models;

- Understand and apply geometric properties and relationships;

- Develop an appreciation of geometry as a means of describing the physical world.

# University of Illinois Committee on School Mathematics (UICSM)

## EXERCISES

**Part A**

Nancy took eight pictures of her brother's new bicycle.

Here is one of the pictures.

Paul, her brother, wanted this particular picture for himself, so Nancy told him to look through the eight negatives for the correct one and she would get a reprint of it. Which of the following is the correct negative ?

*Discuss what was "wrong" with the rejected negatives.*

## Angles and Measures

"Angles and Measures" helps students understand and apply geometric properties and relationships, and develops an appreciation of geometry as a means of describing the physical world.

The exercise on this page comes from a chapter which introduces students to the measurement of turns, or angles, in degrees. They learn to use protractors to measure these angles in the exercises provided. Students discover whether two angles are congruent by using tracings, and through their understanding of the basic motions of flips, slides, and turns. In structured sequences they are introduced to the concept of mapping which, with suitable restrictions, leads them to the concept of congruence. Considerable practice in using tracings to compare figures has been built into the text to help students become aware of the basic properties of congruence.

- For further information about UICSM turn to page 118.

From UICSM, <u>Motion Geometry: Congruence</u>: "Angles and Measures," page 70-76. The UICSM materials were developed at the University of Illinois. For further information contact The Learning Team.

*Measurement activities are a crucial way for students to explore and interpret their world. These lessons illustrate measurement as a significant connection between mathematical topics and the usefulness of mathematics and measurement in everyday situations.*

# Teaching Integrated Mathematics and Science (TIMS)

## View Tube

"View Tube" uses measurement to describe and compare phenomena.

Although the "View Tube" experiment is simple in itself, it contains quantitative elements including interpolation, extrapolation, proportional reasoning, controlling variables and inductive logic. The students look at a meter stick through the view tube (a cardboard cylinder, such as a toilet paper roll) and collect and graph data to find the relationship between two variables: the distance from the viewer to the meter stick and the length of the part of the meter stick they can see. Once they have found that the two variables are proportional, they can use the view tube as a range finder or a height finder.

Teacher lab instructions, teacher lab discussions, experiment pages, and data tables are all provided.

- For further information about TIMS turn to pages 2, 22, 47 and 114.

From TIMS, Experiment 204: "View Tube." TIMS materials were created at, and are available from, the project office at the University of Illinois at Chicago.

# View Tube

## Teacher Lab Discussion

The **View Tube** experiment is simplicity itself. All the student has to do is look through a tube at a meter stick. Yet the experiment contains all the TIMS quantitative elements: interpolation, extrapolation, proportional reasoning, controlling variables, and inductive logic. The experiment also has marvelous applications, from learning how the Egyptians built their pyramids to measuring the size of a giraffe at the zoo. We can use the view tube as a range finder if we are lost or to measure the height of our school building.

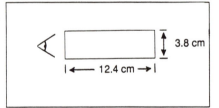

Fig. 1

Sticking to the maxim that cheap is best, the view tube can be the cardboard cylinder of a toilet paper roll. Our tube is illustrated in **Fig. 1**. The dimension will vary from brand to brand of toilet paper. That's okay.

For the **View Tube** experiment a meter stick is

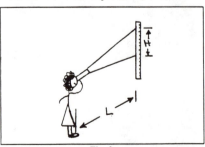

Fig. 2

taped to the wall and the student looks at the meter stick through the tube. Depending upon where the student stands, he or she will see different lengths of the meter stick. The manipulated variable is the distance from the viewer's eye to the meter stick, L, and the responding variable is the amount of the meter stick seen through the field of view, H. This is illustrated in **Fig. 2**.

If you want to force the children into some subtraction, you can make them aim the view tube at the center of the ruler, and then have them subtract the numbers that are visible at the top and bottom to find H (**Fig. 3a**). Most children like to raise the tube so that the top corresponds to the top of the ruler and then find H by reading the number at the bottom

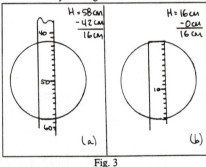

Fig. 3

(**Fig. 3b**). One word of caution: the viewer must keep the tube the same distance from his or her eye for each measurement. Because of this, trading jobs is not a good idea. One partner may have glasses and the other may not; one may have deep set eyes and the other may not. In either case they will measure different H's for the same L.

It is very important to measure L from the <u>viewer's eye</u>. But we do not want children poking around each other's eyes. It turns out, fortunately, that your eyes line up with the instep of your foot,

**The *Standards* state that in grades 5–8 the mathematics curriculum should include extensive concrete experiences using measurement so that students can–**

- Extend their understanding of the process of measurement;

- Estimate, make and use measurements to describe and compare phenomena;

- Select appropriate units and tools to measure to the degree of accuracy required in a particular situation;

- Understand the structure and use of systems of measurement;

- Extend their understanding of the concepts of perimeter, area, volume, angle measure, capacity, weight and mass;

- Develop the concepts of rates and other derived and indirect measurements;

- Develop formulas and procedures for determining measures to solve problems.

# Unified Science and Mathematics for Elementary Schools  (USMES)

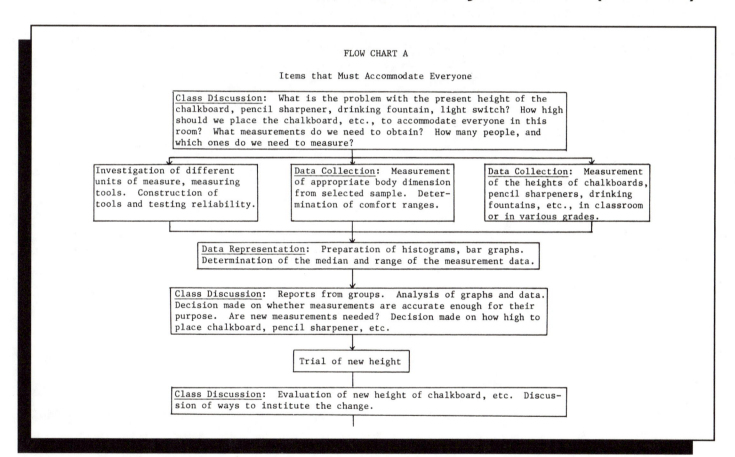

FLOW CHART A

Items that Must Accommodate Everyone

**Class Discussion:**  What is the problem with the present height of the chalkboard, pencil sharpener, drinking fountain, light switch?  How high should we place the chalkboard, etc., to accommodate everyone in this room?  What measurements do we need to obtain?  How many people, and which ones do we need to measure?

**Investigation of different units of measure, measuring tools.  Construction of tools and testing reliability.**

**Data Collection:**  Measurement of appropriate body dimension from selected sample.  Determination of comfort ranges.

**Data Collection:**  Measurement of the heights of chalkboards, pencil sharpeners, drinking fountains, etc., in classroom or in various grades.

**Data Representation:**  Preparation of histograms, bar graphs. Determination of the median and range of the measurement data.

**Class Discussion:**  Reports from groups.  Analysis of graphs and data. Decision made on whether measurements are accurate enough for their purpose.  Are new measurements needed?  Decision made on how high to place chalkboard, pencil sharpener, etc.

**Trial of new height**

**Class Discussion:**  Evaluation of new height of chalkboard, etc.  Discussion of ways to institute the change.

## Designing for Human Proportions: Teacher Resource Book

"Designing for Human Proportions" helps students extend their understanding of the process of measurement.

In this unit, students are challenged to design or make changes in things they use or wear.  After an initial observation in the classroom, the class may identify several objects that do not fit properly or

are the wrong height for some in the classroom,  such as the drinking fountain, chalkboard, desks and smocks.  To determine the proper heights and sizes for a varied population, the students become involved with many aspects of measurement to help in the problem-solving process. The data they collect is graphed and analyzed. To see the spread of measurements, students put the data on a histogram. For example, they may try to decide how high to place a chalkboard and then decide to find the median reach of the class, determining at the same time

how different the reach of some students is from the median.  At the end of the unit, students make a prototype to test their ideas and submit them to the appropriate authorities as recommendations for change.

- For further information about USMES turn to pages 3, 31 and 115.

From USMES, "Designing for Human Proportions." The materials were created at the Education Development Center and for further information contact The Learning Team.

# Section I The *Standards* and Examples for Their Application

## Grade level 9—12

*All students can profit from discussions of specific problem-solving techniques and instructional settings that encourage investigation, cooperation and communication. Perfecting these skills helps answer questions raised in everyday life and other academic disciplines, as well as further extend students' mathematical thinking. The lessons below introduce advanced problem-solving strategies and reinforce processes already learned.*

# Sourcebook of Applications of School Mathematics

## Extending Some Sports Applications

Lessons about sports can be highly motivating. This unit helps students recognize, formulate and investigate sports problems with mathematical content. It helps to increase their confidence in applying problem-solving strategies to problems outside of mathematics.

"The Mile Record" asks students to find an approximate linear relationship between the mile record and the year, as detailed in a table and graph provided. As an extension to this problem-solving activity, additional questions are also provided, as shown on this page. These extending techniques may include generalizing the situation, localizing the data, comparing different events, looking at similar problems, and computer simulations.

• For further information about the *Sourcebook of Applications* turn to pages 75 and 113.

From Sourcebook of Applications of School Mathematics, "Extending Some of the Sports Applications in this Volume," pages 293-296. Prepared by a Joint Committee of the MAA and the NCTM and available from NCTM.

---

*Some Project Problems*

7.D    EXTENDING SOME OF THE SPORTS APPLICATIONS IN THIS VOLUME

It is almost universally believed that sports problems are highly motivating. In the past, this motivation was probably more to boys than to girls, but the recent growth of women's sports has narrowed the motivational gap.

Although a teacher might wish to use sports situations purely as a motivation, some problems lend themselves to very good mathematics and interesting application. In this short essay, we suggest how three of the earlier problems in the volume can be extended or expanded.

1.    The mile record    (Chapter 2, problem 121)

The original problem asks the student to find an approximate linear relationship between the mile record and the year. This is the kind of problem found in linear regression theory. Here are some possible extensions.

Is the situation unusual? The mile record chronology seems to be unusual because the records lie so close to a line. Find out how the records for other distances have changed with time. Runners World is a magazine which may be helpful here. It turns out that more are linear than one would expect.

Does the situation hold for other sports? Try swimming. Have the records there decreased linearly? Or try auto racing. Has the Indianapolis 500 average speed increase linearly? (Or pick another race.)

What is the best possible? Here one can speculate. Should the best possible mile record be four times the best possible quarter-mile record? Esquire magazine gave some predictions in the July, 1976, issue.

Localize the data. Does the pattern of records for your school or your state follow the world pattern? (e.g., is it linear? is it decreasing as quickly?)

Compare different events. How does the evolution of the mile record compare with that for the two-mile? Or how does the men's mile record compare with that for women? (Here it may only be possible to work with other

The *Standards* state that in grades 9–12 the mathematics curriculum should include the refinement and extension of methods of mathematical problem solving so that all students can–

• Use, with increasing confidence, problem-solving approaches to investigate and understand mathematical content;

• Apply integrated mathematical problem-solving strategies to solve problems from within and outside mathematics;

• Recognize and formulate problems from situations within and outside mathematics;

• Apply the process of mathematical modeling to real-world problem situations.

# Secondary School Mathematics Curriculum Improvement Study (SSMCIS)

Optimal Routing.  One of the problems faced by an industrial waste removal company was the routing of their trucks to make pickups at various companies served.  The dispatcher had 7 locations marked on the city street map as below.  He copied the map with directions for the driver as indicated.  He thought he had done an efficient job of mapping out the pickup route.  Do you agree?

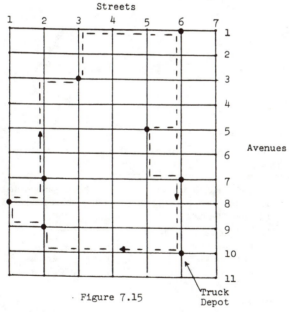

Figure 7.15

The map (Figure 7.15) is a model of the street grid in which the waste removal truck will operate.  You recognize it as a lattice digram of {1,2,3,4,5,6,7} x {1,2,3,4,5,6,7,8,9,10,11}.  The route is a polygon; you can check that it is the shortest possible path touching all dotted vertices.  But, as the dispatcher heard at the end of the day, the indicated route was far from ideal.  A partial list of the driver's complaints follows:

    (1)  First, third, and ninth avenues are one-way in the direction opposite to the indicated route.

    (2)  Fifth street is closed to trucks.

## Modeling and Simulation

"Just as an architect's blueprints model a house, or a miniature railroad models the Penn-Central, the structures of mathematics model a variety of 'real-life' situations.  The models are not identical with the physical reality, but there is an intimate correspondence.  Deductions in one system correlate to facts in the other." (*Unified Modern Mathematics*.)

SSMCIS materials help students recognize and formulate problems from situations outside of mathematics and apply the process of mathematical modeling to real-world situations.

The examples given in this lesson help students to recognize, construct and analyze mathematical models.  Students analyze two models: the first is based on an optimal route for an industrial waste removal truck and the second examines the rate at which objects fall to earth, i.e. velocity in free-fall.  The students discuss instances when models fail and examine the limitations of mathematical modeling.  Most of the exercises are open-ended and include evaluation of the assumptions behind students' models and possible limitations to their validity.

• For further information about SSMCIS turn to pages 63 and 112.

From Unified Modern Mathematics, Book VI: "Modeling and Simulation," pages 380-391. SSMCIS materials were developed by the Secondary School Mathematics Curriculum Improvement Study at Teachers College and were originally published by Teachers College Press.  For further information contact The Learning Team.

*Active student participation in learning can provide many opportunities for discussing, questioning and summarizing mathematical ideas, thus enhancing the acquisition of mathematical concepts. These lessons reinforce the understanding of connections between mathematics and our society.*

## The Language of Functions and Graphs

### Are Graphs Just Pictures?

"Are Graphs Just Pictures?" helps students clarify their thinking about mathematical ideas and relationships, as well as express those ideas orally and in writing.

This lesson is designed to expose and provoke a discussion about the common misconception that graphs are mere "pictures" of an underlying situation rather than abstract representations of relationships. Students are encouraged to discuss the possible errors which may result when interpreting graphs. In the problem shown on this page, for example, students may feel that since the path of the ball goes "up and down," the graph should also go "up and down." Discussions in which the students are made aware of the possible inconsistencies in their beliefs are suggested.

• For further information about The Language of Functions and Graphs turn to pages 37, 44, 67, 69 and 100.

From Shell, <u>The Language of Functions and Graphs</u>, Unit A2: "Are Graphs Just Pictures?" pages 74-81. <u>The Language of Functions and Graphs</u> materials were developed by, and are available from, the Shell Centre for Mathematical Education at the University of Nottingham in the United Kingdom.

**A2   ARE GRAPHS JUST PICTURES?**

**Golf Shot**

How does the speed of the ball change as it flies through the air in this amazing golf shot?

* Discuss this situation with your neighbour, and write down a clear description stating how you both think the speed of the golf ball changes.

* Now sketch a rough graph to illustrate your description:

Speed of the ball

Time after the ball is hit by the golf club.

IS THAT A PICTURE OF THE MOUNTAIN YOU FELL OFF?

The *Standards* state that in grades 9–12 the mathematics curriculum should include the continued development of language and symbolism to communicate mathematical ideas so that all students can–

• Reflect upon and clarify their thinking about mathematical ideas and relationships;

• Formulate mathematical definitions and express generalizations discovered through investigation;

• Express mathematical ideas orally and in writing;

• Read written presentations of mathematics with understanding;

• Ask clarifying and extending questions related to mathematics they have read or heard about;

• Appreciate the economy, power and elegance of mathematical notation and its role in the development of mathematical ideas.

# Contemporary Applied Mathematics (CAM)

## A. Football Statistics

We now employ circle-and-ray glyphs to examine much different data: the offensive attributes of ten National Football League clubs. We depict the characteristics and effectiveness of these offenses.

The data, from the Official 1979 NFL Statistics Guide, is mapped onto a glyph as follows:

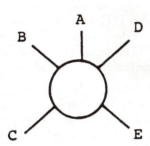

Ray *A*: Represents the number of punts; the longer the ray, the fewer the punts.

Ray *B*: The average number of yards per rush (run) attempt.

Ray *C*: The number of touchdowns scored on running plays.

Ray *D*: The number of completed passes.

Ray *E*: The number of touchdowns scored by passing.

Note that longer rays are always preferable to shorter ones.
Here are two sample glyphs:

## Graphs of Multivariate Data

This lesson is an excellent example of the economy, power and elegance of mathematical notations, and their role in the development of mathematical ideas.

Glyphs, a form of modern picture writing, are a valuable tool for visualizing data based on many variables. In this example each line represents a variable. In Activity A, pictured here, the data are concerned with football statistics. Ray A, for example, represents the number of punts; the longer the ray the fewer the punts, etc. In these activities, students use the glyphs to analyze patterns and answer questions such as: Which team, Dallas or Baltimore, scored more points by rushing? Which team punted more often? The glyph representing Baltimore has no Ray B, i.e the length of Ray B is zero. Does this mean that Baltimore averaged zero yards per attempt?

This lesson introduces students to a variety of glyphs and their varied applications.

• For further information about CAM turn to pages 65, 78, and 92.

From Contemporary Applied Mathematics, Glyphs, Part IV, "Graphs of Multivariate Data," pages 8-13. CAM materials were authored by Wayne Copes, William Sacco, Clifford Sloyer and Robert Stark, and are available from Janson Publications, Inc.

*Formulating conjectures from a pattern of observations and testing those conjectures by logical verification are essential for the development of reasoning skills. These lessons help students experience and expand the role of inductive and deductive reasoning in and outside the mathematical setting. They also encourage the use of appropriate validating mathematical proofs.*

# Discovering Geometry

## Inductive Reasoning

Inductive reasoning is the process of observing data, recognizing patterns, and making generalizations from the data. Such generalizations are called conjectures. This lesson helps students make and test conjectures, formulate counter examples, and follow logical arguments.

This lesson consists of several exercise sets. In the exercise on this page, students are asked to make conjectures using inductive reasoning. In a second exercise they are asked to make careful observations based on a series of pictures. In a third, they are given scenarios with incorrect conjectures and are asked to find the mistakes. In the final exercise, they must think of an example in which inductive reasoning is used correctly and one in which it is used incorrectly.

The Textbook includes a Teacher's Guide and Answer Key, a Teacher's Resource Book and several software disks.

• For further information about *Discovering Geometry* turn to pages 70 and 95.

From Discovering Geometry: An Inductive Approach: Chapter 1, "Inductive Reasoning," pages 39-69. Discovering Geometry is available from Key Curriculum Press.

### EXERCISE SET A

Use inductive reasoning to make a conjecture.

1. Caveperson Stony Grok picks up a rock, drops it into a lake, and watches it sink. He picks up a second rock, drops it into the lake, and it also sinks. He does this five more times, and each time the rock heads straight to the bottom of the lake. Stony conjectures: "Ura nok seblu," which translates to: —?—.

2. A mathematician lands at the airport of the kingdom of Moravia. He desperately needs to use the bathroom, but he is very shy, and the social customs of the kingdom would not permit him to use the wrong bathroom. He locates the doors to what appear to be two bathrooms. He observes men enter the door marked "Warvan" and women enter the door marked "Cupore." He is finally ready to make his conjecture. How does he spell relief?

3. Salesperson Henrietta Cluck is selling square-egg makers door to door. At the first house she tells a joke about robot chicken eggs, gets a laugh, and sells a square-egg maker. At the second house she uses a "creative cooking technique" approach. Her approach is informative, but she doesn't sell her square-egg maker. At the third, fourth, sixth, and eighth houses she tries the robot egg joke, and each time she sells the square-egg maker. At the fifth and seventh houses she tries her "creative cooking techniques" approach and is unsuccessful each time. Henrietta conjectures: —?—.

**Henrietta Cluck Sells Square Egg Makers**

4. Juan went to a restaurant and ate sushi for the first time. A few hours later he broke out in a terrible rash. Two days later he went to another restaurant, ate sushi, and broke out in a rash. A week later Juan decided to give sushi one more try, but, unfortunately, he developed the same irritating rash. Juan conjectures: —?—.

The *Standards* state that in grades 9–12 the mathematics curriculum should include numerous and varied experiences that reinforce and extend logical reasoning skills so that all students can–

- Make and test conjectures;

- Formulate counter examples;

- Follow logical arguments;

- Judge the validity of arguments;

- Construct simple valid arguments;

and so that, in addition, college-intending students can–

- Construct proofs for mathematical assertions, including indirect proofs and proofs by mathematical induction.

# Secondary School Mathematics Curriculum Improvement Study (SSMCIS)

1.4  Connectives:  And, Or

The compound mathematical sentence

5 is prime and 5 is a multiple of 3,

is composed of two simple statements joined by the connective "and."  Is the compound sentence a statement?  Is it true or false?

Clearly the statement "5 is prime" is true and "5 is a multiple of 3" is false.  It seems reasonable that for the compound sentence to be true <u>both</u> parts must be true.  According to this, then the given sentence is false (but is a statement).  What probably suggested that <u>both</u> parts had to be true if the whole sentence is considered to be true is the word "and" connecting the two parts.  In this case, mathematics agrees with intuition.  Statements of the form "P and Q" are true if <u>both</u> P and Q are true and false if either one of them or both, are false.

Example 1.  The compound statement "5<3 and 2>7" is false since neither "5<3" nor "2>7" is true.

Example 2.  The compound statement S: "2 + 3 = 5 and 4 ≠ 7" is true.  Since "2 + 3 = 5" is true and "4 ≠ 7" is true, both parts of S are true and therefore S is true.

Example 3.  S: "x is a whole number greater than 5 and x is a whole number less than 8."  Of course, this is not a statement--it is an open

## Mathematical Language

"Mathematical Language" helps students judge the validity of their arguments.

Students are introduced to the language used in formal mathematical proofs.  Words such as "and" and "or" are defined, using truth tables, and the mathematical definitions are compared with ordinary English usage.  Truth tables are also given for "if-then," "for all," "there exists," and other types of mathematical statements.

Using truth tables, statements from algebra and geometry are analyzed to illustrate their logic as well as to determine their truth or falsity.

This lesson not only helps students determine whether arguments are true or false but also helps them to further explore mathematical logic.

- For further information about SSMCIS turn to pages 59 and 112.

From Unified Modern Mathematics, Course II, Part 1, Chapter 1: "Mathematical Language and Proof," pages 1-36.  SSMCIS materials were developed by the Secondary School Mathematics Curriculum Improvement Study at Teachers College and were originally published by Teachers College Press.  For further information contact The Learning Team.

# 9–12 Standard 4: Mathematical Connections

*It is essential to understand the connections between mathematics and problem situations that arise in the real world as well as between mathematics and other academic disciplines. It is also important to be able to translate, and therefore to make connections between, different mathematical representations of the same problem. The lessons below help the student develop insight into, and procedures for making, mathematical connections.*

## Introduction to College Mathematics (ICM)

### Models and Mathematics: Geometric Probability

This lesson helps students apply the process of mathematical modeling to real-world situations. A variety of models, such as graphs and equations, illustrate the connection between different mathematical representations.

The lesson begins with a general overview of mathematical modeling and an explanation of why models created from data are useful tools in problem solving.

In the "Tape Recorder Problem," shown here, a variety of models are created to help determine the probability of where a taped conversation was erased. Additional probability problems in the lesson examine the strengths and limitations of road maps and Newtonian physics.

- For further information about ICM turn to pages 77 and 97.

From Introduction to College Mathematics, <u>Contemporary Precalculus through Applications</u>: "Models and Mathematics: Geometric Probability," pages 3-4, 17-23. The program was developed by the Department of Mathematics and Computer Sciences, North Carolina School of Science and Mathematics, and is available from Janson Publications.

### 9.1 The Tape Recorder Problem

A tape recording is made of a meeting between two businessmen. Their conversation starts at the 21st minute on the tape, and lasts only 8 minutes. The tape will record for 60 minutes. While playing back the tape one of the businessmen accidentally erases 15 minutes of the tape, but he doesn't know where.

1. Find the probability that the entire conversation was erased.

2. Find the probability that some part of the conversation was erased.

3. Suppose the exact position of the conversation on the tape is not known, except that it began sometime after the 21st minute. Find the probability that the entire conversation was erased.

In part 1, a single random event occurs — that is, the starting time of the 15-minute erasure. If we let $x$ be the number of minutes from the beginning of the tape to the start of the erasure, then $x$ can vary from 0 to 45. (Do you see why $x$ cannot take on values between 45 and 60?) Using a number line to represent the tape, we can see that the sample space is 45 units long. An illustration is shown in Figure 24.

Where does the event occur? The conversation lasts from the 21st minute to the 29th minute, so the erasure can start as early as the 14th minute and as late as the 21st minute in order to erase the entire conversation. Thus, the value of $x$ must be between 14 and 21 in order to be in the event space. The event space is 7 units long, the sample space is 45 units long, and the probability is 7/45, or about 0.16.

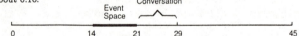

Figure 24: Tape Recorder Problem – Part 1

For part 2, we can still let $x$ represent the starting time of the erasure. The sample space is the same as it was in part 1, but the event space is larger. The event space now includes values of $x$ that result in any portion of the conversation being erased. In order to find the event space, the situation can be analyzed algebraically, or you can use trial and error. You should find that $x$ can be anywhere between 6 minutes and 29 minutes. So the event space is 23 units long, and the probability is 23/45, or about 0.51.

In part 3, two random events occur — one is the starting time of the erasure and the second is the starting time of the conversation. Let $x$ represent the starting time of the erasure, and let $y$ be the starting time of the conversation. Since the erasure is 15 minutes long, $x$ can vary from 0 to 45. Since the conversation is 8 minutes long, $y$ can vary from 21 to 52. The sample space consists of all pairs $(x, y)$ in the rectangle in Figure 25.

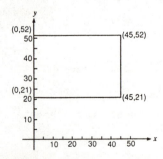

The *Standards* state that in grades 9–12 the mathematics curriculum should include investigations of the connections and interplay among various mathematical topics and their applications so that all students can–

- Recognize equivalent representations of the same concept;

- Relate procedures in one representation to procedures in an equivalent representation;

- Use and value the connections among mathematical topics;

- Use and value the connections between mathematics and other disciplines.

# Contemporary Applied Mathematics (CAM)

## Reachability

With "Reachability" students can use and appreciate the connections among mathematical topics as well as the connections between mathematics and other disciplines.

In this lesson, students use directed graphs to model a communications network between a set of cities in order to find if a message from each vertex can go to any other. To solve this problem, they must learn to manipulate the adjacency matrices for their graphs and raise these matrices to various powers in order to determine the number of paths between the vertices in the graph. Students then determine which other vertices can be reached from each vertex.

- For further information about CAM turn to pages 61, 78 and 92.

To illustrate reachability, suppose a directed communications network exists among a set of cities. (In this problem, two-way communication is not always possible because of transmitter and receiver difficulties between cities.) The di-graph for the situation is:

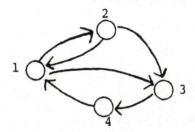

An interesting question which arises in situations like this is: Is it possible to get a message from each vertex to any other? That is, is every vertex "reachable" from every other? We shall see that the answer is obtainable by manipulating the adjacency matrix of the di-graph and then we shall explore some additional applications of reachability.

**74.** Fill in the adjacency matrix, $M$, below, for the communication problem.

|  | To | | | |
|---|---|---|---|---|
| **From** | 1 | 2 | 3 | 4 |
| 1 | | | | |
| 2 | | | | |
| 3 | | | | |
| 4 | | | | |

Let's review the definition of $M$'s elements. The element in the $i$th row and $j$th column is "1" if there is a directed edge going *from* the $i$th vertex *to* the $j$th vertex. Otherwise, the entry is a zero. Interpreted another way, that element is "1" if there is a *one*-edged path *from* vertex $i$ *to* vertex $j$.

Consider what happens when the adjacency matrix is multiplied by itself.

From the rules for matrix multiplication, the $(i, j)$th entry of the resulting matrix, $M^2$, is formed by the *inner product* of the $i$th row and $j$th column of the original matrix, $M$. For example, in the diagram below, the element of $M^2$, in the 1st row and 4th column is found by multiplying each entry of the 1st row by the corresponding entry in the 4th column of $M$ and adding the products. The result is $M^2(1, 4) = 1$.

From Contemporary Applied Mathematics, Graph Theory, Part IV-B: "Reachability," pages 47-51. CAM materials were authored by Wayne Copes, William Sacco, Clifford Sloyer and Robert Stark, and are available from Janson Publications, Inc.

*Algebraic representation is used to communicate a great deal of mathematics. It is also a process that encourages generalized abstract thinking and insights. The two lessons below help students see how algebra serves as a problem-solving tool.*

## Lane County Mathematics Project

### The Tower of Brahma

"The Tower of Brahma" illustrates many aspects of the *Standards*. Through this lesson, students learn to appreciate the power of algebra, mathematical abstraction and symbolism.

"The Tower of Brahma," shown on this page, challenges students to complete an ancient Hindu puzzle. Students are asked to move a stack of different sized discs from one post to another, subject to four restrictions.

To solve the puzzle, students must apply a variety of problem-solving skills including using a model, making a systematic list, looking for patterns, and making predictions based on patterns. Algebraic concepts are found within the problem, as students must generalize their results into a recursive formula for the number of moves necessary for any stack.

- For further information about the Lane County Mathematics Project turn to pages 6, 17 and 99.

From the Lane County Mathematics Project, Problem Solving in Mathematics, Grade 9, Algebra: "Tower of Brahma," pages 285-292. The Problem Solving in Mathematics series was created by the Lane Education Service District and is available from Dale Seymour Publications.

---

#### THE TOWER OF BRAHMA

An ancient Hindu legend goes like this.

Brahma placed 64 disks of gold—each one a different size—in a stack so the largest disk was on the bottom. The temple priests were told to transfer the disks according to the following rules:

a. Only three stacks can be used.

b. Only one disk at a time can be moved.

c. No disk may be placed on top of a smaller disk.

d. Use the fewest moves possible.

The legend states that the world would vanish when the original stack of 64 disks was transferred to one of the other two stacks.

1. Guess the number of moves you think the priests would have to make. _____

2. Use the disks and playing board on the next page. Find the fewest number of moves to transfer a stack with 1 disk, 2 disks, 3 disks, etc. Complete the table.

3. How many years would it take for the priests to transfer a stack of 32 disks if they made one move per second?

| Number in the stack | Fewest Moves |
|---|---|
| 1 | |
| 2 | |
| 3 | |
| 4 | |
| 5 | |
| . | |
| . | |
| . | |
| 10 | |
| . | |
| . | |
| . | |
| $n$ | |

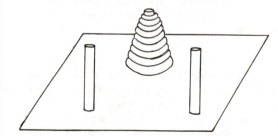

# The Language of Functions and Graphs

## CARBON DATING

Carbon dating is a technique for discovering the age of an ancient object, (such as a bone or a piece of furniture) by measuring the amount of Carbon 14 that it contains.

While plants and animals are alive, their Carbon 14 content remains constant, but when they die it decreases to radioactive decay.

The amount, *a*, of Carbon 14 in an object *t* thousand years after it dies is given by the formula:

$$a = 15.3 \times 0.886^{\,t}$$

(The quantity "*a*" measures the rate of Carbon 14 atom disintegrations and this is measured in "counts per minute per gram of carbon (cpm)")

1   Imagine that you have two samples of wood. One was taken from a fresh tree and the other was taken from a charcoal sample found at Stonehenge and is 4000 years old.

   How much Carbon 14 does each sample contain? (Answer in cpm's)

   How long does it take for the amount of Carbon 14 in each sample to be halved?

   These two answers should be the same, (Why?) and this is called the *half-life* of Carbon 14.

2   Charcoal from the famous Lascaux Cave in France gave a count of 2.34 cpm. Estimate the date of formation of the charcoal and give a date to the paintings found in the cave.

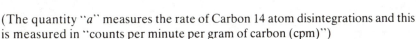
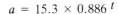
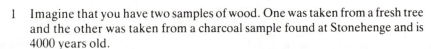

3   Bones A and B are *x* and *y* thousand years old respectively. Bone A contains three times as much Carbon 14 as bone B.

   What can you say about *x* and *y*?

## Carbon Dating

In this lesson, students are asked to create tables and graphs showing how the amount of carbon-14 in an object (e.g., a bone or a piece of wood) varies over hundreds of years. Through this process, students practice using graphs to interpret expressions, equations and inequalities.

The lesson begins with an explanation of the carbon-dating process, a look at the meaning of the term "half-life," and an example of how an archeological find may be dated. Students then work in small groups to create and analyze graphs to solve a variety of questions. Their solutions involve an exponential equation where various function values are given.

- For further information about *The Language of Functions and Graphs* turn to pages 37, 44, 60, 69 and 100.

From Shell, The Language of Functions and Graphs, "Carbon Dating," pages 170-173. The Language of Functions and Graphs materials were developed by, and are available from, the Shell Centre for Mathematical Education at the University of Nottingham in the United Kingdom.

*The mathematical concept of function, the special correspondences between the elements of two sets, is a unifying idea throughout the mathematics curriculum. These lessons illustrate the power of functions to simplify complex situations and translate real-world problems into mathematical representations.*

# Ohio State University Calculator and Computer Precalculus Project (C2PC)

## Quadratic Functions and Geometric Transformation

In "Quadratic Functions and Geometric Transformation" students learn to analyze the effects of parameter changes on the graphs of functions.

The lesson illustrates how the graph of any function of the form $f(x) = ax^2 + bx + c$ can be obtained from the graph of $y = x^2$ by applying a sequence of geometric transformations. Three of the transformations are rigid-motion transformations that produce congruent graphs. The other transformation stretches or shrinks the graph, producing a new graph. These geometric transformations are also applied to other functions in the lesson.

Use of a graphing calculator or graphing software is assumed for all the lessons in *Precalculus Mathematics: A Graphing Approach.*

- For further information about Ohio State University Calculator and Computer Precalculus Project turn to pages 82, 84 and 107.

From Ohio State University Calculator and Computer Precalculus Project, Precalculus Mathematics: A Graphing Approach, Section 3.3, "Quadratic Functions & Geometric Transformation," pages 177-189. The materials were the result of a collaboration between Ohio State University and three Ohio school districts, and are available from Addison-Wesley Publishing Company.

- DEFINITION   A function $f$ is called a **quadratic function** if it can be written in the form $f(x) = ax^2 + bx + c$, where $a$, $b$, and $c$ are real numbers and $a \neq 0$.

  Notice that a quadratic function is a polynomial of degree 2.

- EXAMPLE 1: Draw complete graphs of the functions $y = x^2$, $y = 3x^2$, and $y = \frac{1}{2}x^2$ on the same coordinate system.

  SOLUTION: The three graphs are shown in Figure 3.3.1 and are complete.

  Each graph in Figure 3.3.1 passes through the point $(0,0)$. Excluding the origin, the graph of $y = 3x^2$ lies above the graph of $y = x^2$, and the graph of $y = \frac{1}{2}x^2$ lies below the graph of $y = x^2$. This happens because, for each nonzero real number $x$, the value of $3x^2$ is greater than the value of $x^2$, and the value of $\frac{1}{2}x^2$ is less than the value of $x^2$. In fact, for $x \neq 0$, the graph of $y = ax^2$ will always lie above the graph of $y = x^2$ if $a > 1$, or below the graph of $y = x^2$ if $0 < a < 1$.

- DEFINITION   If $a > 1$, the graph of $y = ax^2$ is said to be obtained from the graph of $y = x^2$ by **vertically stretching** the graph of $y = x^2$ by the factor $a$. If $0 < a < 1$, the graph of $y = ax^2$ is said to be obtained from the graph of $y = x^2$ by **vertically shrinking** the graph of $y = x^2$ by the factor $a$. Vertically stretching or shrinking a given graph by the positive factor $a$, with $a \neq 1$, does not produce a congruent graph.

- EXAMPLE 2: Draw complete graphs of the functions $y = -x^2$, $y = -3x^2$, and $y = -\frac{1}{2}x^2$ on the same coordinate system.

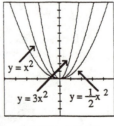

$[-5, 5]$ by $[-5, 10]$

Figure 3.3.1

The *Standards* state that in grades 9–12 the mathematics curriculum should include the continued study of functions so that all students can–

• Model real-world phenomena with a variety of functions;

• Represent and analyze relationships using tables, verbal rules, equations and graphs;

• Translate among tabular, symbolic and graphical representations of functions;

• Recognize that a variety of problem situations can be modeled by the same type of functions;

• Analyze the effects of parameter changes on the graphs of functions;

**and so that, in addition, college-intending students can–**

• Understand operations on, and the general properties and behavior of, classes of functions.

# The Language of Functions and Graphs

Imagine that a doctor prescribed a drug called Triazolam. (Halcion*).
After taking some pills, the drug eventually reaches a level* of $4\mu g/l$ in the blood plasma.
How quickly will the drug wear off?

Look at the table shown below:

| Drug name (and Brand name) | Approximate formula |
|---|---|
| Triazolam (Halcion*) | $y = A \times (0.84)^x$ |
| Nitrazepam (Mogadon*) | $y = A \times (0.97)^x$ |
| Pentobombitone (Sonitan*) | $y = A \times (1.15)^x$ |
| Methohexitone (Brietal*) | $y = A \times (0.5)^x$ |

KEY  A = size of the initial dose in the blood

y = amount of drug in the blood

x = time in hours after the drug reaches the blood.

For Triazolam, the formula is $y = A \times (0.84)^x$

In our problem the initial dose is 4 $\mu g/l$, so this becomes

$$y = 4 \times (0.84)^x$$

---

* Please note that in this worksheet, doses and blood concentrations are not the same as those used in clinical practice, and the formulae may vary coniderably owing to physiological differences between patients.

## Looking at Exponential Functions

"Looking at Exponential Functions" provides a medical context within which the properties of exponential functions may be discussed.

The lesson examines a number of issues related to the effects of medication on patients. For example, as shown on this page, in order for doctors to make informed decisions concerning a prescribed drug, they must examine many interrelated factors, such as the size of the initial dose in the blood, the amount of drug in the blood, and the rate at which the drug wears off in the blood stream. They must also be able to calculate the effects of a certain dose of the drug taken at regular intervals.

This lesson helps students understand operations on, and the general properties and behavior of, exponential functions.

• For further information about *The Language of Functions and Graphs* turn to pages 37, 44, 60, 67 and 100.

From Shell, The Language of Functions and Graphs, Unit B3: "Looking at Exponential Functions," pages 120-125. The Language of Functions and Graphs materials were developed by, and are available from, The Shell Centre for Mathematical Education at the University of Nottingham in the United Kingdom.

*Instruction in synthetic geometry should focus on the interplay between inductive and deductive reasoning as well as on development of students' skills in visualization, pictorial representation, and application in the natural, physical and social arenas. These lessons help students understand shapes and their properties with an emphasis on their real-world context.*

# Discovering Geometry

## Racetrack Geometry

In "Racetrack Geometry" students work on a problem using a geometric model. They must apply their knowledge of the properties of circles in order to calculate their answer.

Specifically, students are asked to design an oval four-lane racetrack with a given set of specifications. They must determine the lengths of the straightaways with semicircular ends for each lane, and must know the properties of a circle in order to arrive at the answer.

In order to make a race fair, they must also determine the starting line for each lane of the track so that an 800-meter race can be run in all four lanes.

The Textbook includes a Teacher's Guide and Answer Key, a Teacher's Resource Book and several software disks.

- For further information about *Discovering Geometry* turn to pages 62 and 95.

From <u>Discovering Geometry: An Inductive Approach</u>, Chapter 6, "Arc Length," pages 282-87. <u>Discovering Geometry</u> is available from Key Curriculum Press.

### *Special Project*
### Racetrack Geometry

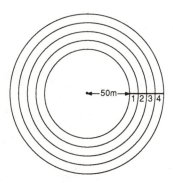

If you had to start and finish at the same line, which lane of the racetrack on the right would you choose to run in? Sure, the inside lane. If the runners in the four lanes were to start and finish at the line shown, the runner in the inside lane would have an obvious advantage because that lane is the shortest. For a race to be fair, runners in the outside lanes must be given head starts.

Your task in this special project is to design a 4-lane oval track with straightaways and semicircular ends. The semicircular ends must have inner diameters of 50 meters so that the distance of one lap in the inner lane is 800 meters. Draw starting and stopping segments in each lane so that an 800 meter race can be run in all four lanes

What do you need to know to design such a track? You will need to determine the length of the straightaways. You will also need to determine the head start for each of the runners in the outer lanes so that each has 800 meters to the finish line. (Use 3.14 for $\pi$.) Before you begin creating your racetrack, you will need to determine:

- Does the radius of the circle play a part in determining the head start?

- Does the width of the lane play a part in determining the head start?

- Does the length of the straightaways play a part in determining the head start?

To answer these questions, try calculating the lengths of a few sample racetracks. For example, if the inner radius of the circular track pictured above is 50 meters and each lane is 1 meter wide, you can calculate the distance each runner must travel in one lap if each runner must stay in his or her own lane.

Copy and complete the table. $r$ is the radius of the circle that defines the inside edge of each lane. $C$ is the circumference. All distances are in meters.

| Lane 1 | Lane 2 | Lane 3 | Lane 4 |
|---|---|---|---|
| $r = 50$ | $r = 51$ | $r = -?-$ | $r = -?-$ |
| $C = 100\pi$ | $C = -?-$ | $C = -?-$ | $C = -?-$ |

To make a race fair, you can look in the table above to determine how much of a head start each runner in the outer lanes must have. For the circular track above, it turns out that the runner in lane 2 must have a head start of $2\pi$ meters over the runner in lane 1; the runner in lane 3 must have a $2\pi$ meters head start over the runner in lane 2; and, the runner in lane 4 must also have a $2\pi$ meters head start over the runner in lane 3. With these head starts, each runner will travel $100\pi$ meters.

The *Standards* state that in grades 9–12 the mathematics curriculum should include the continued study of the geometry of two and three dimensions so that all students can–

- Interpret and draw three-dimensional objects;

- Represent problem situations with geometric models and apply properties of figures;

- Classify figures in terms of congruence and symmetry and apply these relationships;

- Deduce properties of, and relationships between, figures from given assumptions;

**and so that, in addition, college-intending students can–**

- Develop an understanding of an axiomatic system through investigating and comparing various geometries.

# High School Mathematics and Its Applications (HiMAP)

## Two-Dimensional Patterns

"Two-Dimensional Patterns" classifies patterns in terms of congruence and symmetry and helps students learn to apply those relationships.

In this lesson, students learn to identify and mathematically classify two-dimensional patterns. Although there appear to be an infinite number of possibilities, there are actually only seventeen types of two-dimensional patterns. These remarkable patterns have their origins in such varied places as ancient Egyptian paintings and African bark cloth. Students are provided a flow-chart structure based on certain properties of the patterns from which to make an analysis.

The flow-chart analysis of patterns begins by asking for the smallest rotation which the pattern admits, then proceeds to further analyze each pattern through such questions as whether there is a reflection in the pattern or not, which leads to other questions until a type of pattern is established. The flow-chart analysis of patterns is a useful activity for the whole class at once and can arouse most students' interest.

- For further information about HiMAP turn to pages 80 and 96.

From HiMAP Module 4, Symmetry, Rigid Motions and Patterns, Section 5: "Two-Dimensional Patterns," pages 22-30. HiMAP materials were developed by, and are available from, COMAP, Inc.

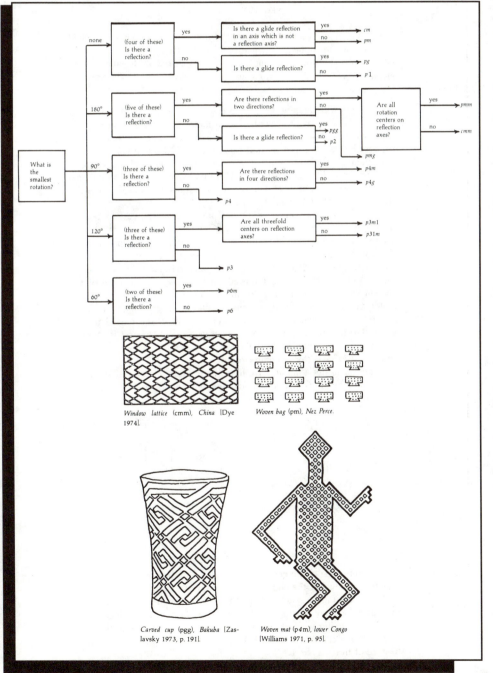

Window lattice (cmm), China [Dye 1974].

Woven bag (pm), Nez Perce.

Carved cup (pgg), Bakuba [Zaslavsky 1973, p. 191].

Woven mat (p4m), lower Congo [Williams 1971, p. 95].

*The important connections between geometry and algebra, and the ability to translate between these two perspectives, is fundamental. Students must learn to compare, contrast and translate among synthetic, coordinate and transformation geometry systems. The two lessons below help develop students' appreciation for varied applications of geometry to real-world problems.*

# School Mathematics Study Group (SMSG)

## Vectors

This lesson introduces vectors, and students are asked to apply them to a variety of problems in physics and mathematics.

"Vectors" begins with an in-depth study of one example: "A man is to cross a river from the left bank to the right. Too lazy to row, he uses a motorboat." Using vectors, students determine his course under several conditions, including:

- His motor fails to start and he drifts down river with the current;
- His motor is working and due to high tide there is no current;
- His motor is working and there is a current.

Vectors are also used to study inclined planes, pulleys and levers. Several applications of Archimedes' Law of the Lever, including using levers to model volumes of solids and multiplication of signed numbers, are discussed.

- For further information about SMSG turn to pages 25 and 111.

From SMSG, <u>Mathematical Methods in Science</u>, Chapter 2, Section 2: "Vectors," pages 58-81. SMSG was developed at the School of Education at Stanford University. For further information contact The Learning Team.

### SECTION 2. VECTORS

The notion of a vector arises quite naturally and is basic to physics and indispensable to applied mathematics. That it is clear from the outset that vectors are good for something makes the topic readily teachable at an elementary level. That vectors are becoming part of the high school program is a real step forward.

We begin with an example. A man is to cross a river from the left bank to the right. Too lazy to row, he uses a motor boat. If his motor fails to start when he casts off, he will drift down river with the tide. Let us suppose him to drift $AB$ in unit time. See Fig. 2.16. If it is high tide so that there is neither a current up nor down river, and his motor is working, he will travel, let us say, $AC$ in unit time. But, if both tide and motor are working, his boat will have velocities due to both. Where will it be at the end of unit time?

The answer comes quite naturally. Consider a special case. A boat at $A$ heading up river at, say, 440 feet per minute (that is 5 m.p.h.) against a down-river current of the same velocity moves neither up nor down river; with both velocities simultaneously it stays put relative to the river bank. At the end of a minute it is in the same position as it would be at the end of two minutes if it moved solely under the influence of the current with no motor for the first minute and under the influence of the motor with no current for the second minute. In the first minute it would move 440 feet down river with the current and in the second minute motor 440 feet back up the (now currentless) river. Thus (at the end of two minutes) it would be in the same position after current and motor acted *successively* (for a minute each) as it would be after both acted *simultaneously* (for a minute). In short, the resultant effect of both forces, current and motor, is that of each acting independently of the other.

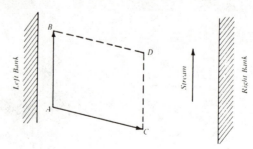

Figure 2.17

Thus, returning to the general case of Fig. 2.16, it is natural to suppose that the boat will at the end of unit time, say, a minute, be at $D$, where $ABDC$ is a parallelogram. See Fig. 2.17. In one minute the boat acted on by current without motor would drift to $B$; in the succeeding minute acted on by motor without current it would go as far as (and in the same direction as) if it started from $A$ instead of $B$, i.e., from $B$ to $D$ (instead of from $A$ to $C$).

- Translate between synthetic and coordinate representations;

- Deduce properties of figures using transformations and using coordinates;

- Identify congruent and similar figures using transformations;

- Analyze properties of Euclidean transformations and relate translations to vectors;

**and so that, in addition, college-intending students can–**

- Deduce properties of figures using vectors;

- Apply transformations, coordinates and vectors in problem solving.

# Madison Project

## Matrices and Space Capsules

"Matrices and Space Capsules" provides an opportunity for students to apply transformations, coordinates and vectors in problem-solving situations.

In this lesson, students predict the motion of a rocket in space by studying its sliding, flipping and rotation transformations through corresponding matrices. The students represent the initial position of a space capsule by a set of points on the Cartesian graph. They operate on those points with a given transformation matrix and find the position of the capsule after the maneuver is completed. They identify matrices which rotate the capsule clockwise and counter-clockwise 90 degrees. Finally, given an initial position and a series of shifts based on these matrices, students predict the final position of the capsule.

- For further information about the Madison Project turn to pages 5, 14, 46, 85 and 101.

From the Madison Project, <u>Explorations in Math</u>, Chapter 41: "Matrices and Space Capsules," pages 372-380. <u>Explorations in Math</u> was written by the Madison Project and is available from Cuisenaire Company of America.

---

STUDENT PAGE 148]    MATRICES AND SPACE CAPSULES    373

Now, it is essential to predict, and to observe, the motion of space capsules very precisely, using appropriate mathematics and high-speed digital computers.

The "flopping" kinds of motions are observed using "before" and "after" pictures of the kind we have just been studying. The "flopping" itself is regarded as a **transformation**, and is studied by means of its corresponding matrix.

[page 148]

(1) Suppose a space capsule is represented by this set of points:

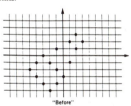

"Before"

The set of points plotted on the graph above is the set

$$\{(2,1), (3,2), (2,3), (1,2), (^-1,1), (^-3,0), (^-4,^-1),$$
$$(^-5,^-2), (^-4,^-3), (^-3,^-4), (^-2,^-5), (^-1,^-4), (0,^-3),$$
$$(1,^-1), (^-2,^-1), (^-1,^-2)\}.$$

Now, at this instant, a computer down on earth sends up a signal which causes the capsule to fire some small "flipping" rockets and "flop over." The computer on earth made a transformation using the matrix

$$\begin{pmatrix} 0 & 1 \\ ^-1 & 0 \end{pmatrix}.$$

Assuming that the rockets all worked correctly, and the capsule did what the computer ordered, what is the new "position" or "attitude" of the space capsule?

(2) Suppose that the capsule started in this position

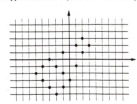

and the computer used the matrix

$$\begin{pmatrix} 0 & ^-1 \\ 1 & 0 \end{pmatrix}.$$

(1) $\begin{pmatrix} 0 & 1 \\ ^-1 & 0 \end{pmatrix} \begin{pmatrix} x_{old} \\ y_{old} \end{pmatrix} = \begin{pmatrix} x_{new} \\ y_{new} \end{pmatrix},$

so

$$x_{new} = y_{old},$$
$$y_{new} = {^\circ}x_{old}.$$

(Alternatively, you could *avoid* the use of variables here, and map one point at a time, using *numbers*:

$$\begin{pmatrix} 0 & 1 \\ ^-1 & 0 \end{pmatrix} \times \begin{pmatrix} 2 \\ 1 \end{pmatrix} = \begin{pmatrix} 1 \\ ^-2 \end{pmatrix},$$

and so on.)

This gives us this set of points:

$$\{(1,^-2), (2,^-3), (3,^-2), (2,^-1), (1,1), (0,3), (^-1,4), (^-2,5),$$
$$(^-3,4), (^-4,3), (^-5,2), (^-4,1), (^-3,0), (^-1,^-1), (^-1,2),$$
$$(^-2,1)\}.$$

If we now plot these points, we get:

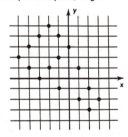

(2) and (3) We suggest you represent the capsule by actual number coordinates, as we did in question 1, and then use the same general method that we used there.

*Many real-world problems rely on the application of trigonometry, i.e. the study of triangle measurement, for their solutions. Trigonometric and circular functions, such as sine and cosine, are fundamental mathematical models for many periodic real-world phenomena. The lessons below help students grasp the basic trigonometric principles and functions.*

## Mathematics Curriculum and Teaching Program (MCTP)

### Trigonometry Walk

"Trigonometry Walk" provides a concrete and physical experience for students to develop a mental image of the sine and the cosine of an angle.

Sine and cosine functions are often difficult concepts for students to understand. By using string to create large triangles on the floor and stepping off the lengths of the sides, students acquire strong mental images of the relationship between the sine and cosine of the angle and the lengths of the opposite and adjacent legs of the triangle. Freed from the constraints of measuring with instruments, students can focus on these relationships and are better able to internalize the experience for later work in the classroom. To complete this lesson, it is important for students to have access to a scientific calculator that has sine, cosine and tangent functions.

- For further information about MCTP turn to pages 10, 16, 32, 51 and 102.

From MCTP Activity Bank: Volume 1, Chapter 4, "Trigonometry Walk," pages 219-224. The Activity Bank materials were assembled by MCTP and are available from the Curriculum Corporation.

### Trigonometry walk

*By children physically experiencing (by walking) the sine, and later the cosine of a triangle, a strong mental image of the relationship between angle and length is created. This cognitive image can then be used in formal situations back in the classroom, that is, the episode can be relived as a concrete basis for understanding sine and cosine.*

**1. Demonstrating the method**

Mark out a ten-pace radius, quarter circle.

WE WILL BE MEASURING LENGTHS BY STEPPING THEM OUT, SO YOUR STEPS MUST BE ALL THE SAME SIZE.
MARK A POINT ON THE SIDE LINE AND TAKE TEN VERY EVEN STEPS.
FROM THAT POINT WE WILL DRAW A QUARTER CIRCLE

KNEEL DOWN, HOLDING STRING

10 PACES

SIDE LINE OF BASKETBALL COURT

**2. What happens to the sides of the triangle as you go along the quarter circle?**

The walker creates right-angled triangles as he or she walks the circle. It is important for watching pupils to visualise these triangles as the angle at the centre changes.

If the walker is stopped once or twice on his or her walk, the triangle could be highlighted with string.

*'Pupils seemed to get a strong visual image that there is an infinite number of triangles, and that each is different.'*

WALKER (holding string)

THIS PUPIL KEEPS STRING TAUT

CHANGING ANGLE

MARK THIS POINT WITH CHALK

PUPIL AT BASE LINE SHUFFLES ALONG, MAINTAINING A RIGHT ANGLE

WATCH THE TRIANGLES THEY CREATE.
ONE SIDE IS ALWAYS THE SAME.
BUT WATCH WHAT HAPPENS TO THE PERPENDICULAR AND THE BASE

THE PERPENDICULAR KEEPS GETTING LONGER

THE BASE KEEPS GETTING SHORTER

The *Standards* state that in grades 9–12 the mathematics curriculum should include the study of trigonometry so that all students can–

- Apply trigonometry to problem situations involving triangles;

- Explore periodic real-world phenomena using the sine and cosine functions;

**and so that, in addition, college-intending students can–**

- Understand the connection between trigonometric and circular functions;

- Use circular functions to model periodic real-world phenomena;

- Apply general graphing techniques to trigonometric functions;

- Solve trigonometric equations and verify trigonometric identities;

- Understand the connections between trigonometric functions and polar coordinates, complex numbers, and series.

# Sourcebook of Applications of School Mathematics

*Trigonometry and Logarithms*

4.8 (Carpentry)

You have a summer job putting up outhouses in county parks. You must pre-cut the rafter ends so that they will be vertical when in place. The front wall is 8' high, the back wall is 6½' and the distance between the walls is 8'. At what angle should you cut the rafters?

4.8 SOLUTION

$\tan \theta = 8 \div \frac{3}{2} = 5 \ 1/3$,

so $\theta = 80°$

(rounded from $79°23'$)

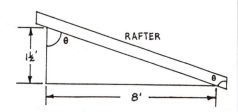

4.9 (Astronautics)

A degree may seem like a very small unit, but an error of one degree in measuring an angle may be very significant. For example, suppose that a laser beam is directed toward the visible center of the Moon and that it misses its assigned target by 30 seconds. How far is it (in miles) from its assigned target? (Take the distance from the surface of the Earth to that of the Moon to be 234,000 miles.)

4.9 SOLUTION

234,000 sin 30" = 234,000 · .00015 = 34 miles.

Comment: Since the radius of the Moon is over 1000 miles, it is safe to treat the part of the Moon's surface involved in the problem as a piece of a plane perpendicular to the laser beam. An interesting variant is obtained by asking about where the beam reflected from a mirror on the Moon at various angles will strike the Earth.

4.10 (Forestry)

A hiker is hiking up a 30° slope and he counts his steps from the base of a tree. After hiking about 80' he estimates that his eyes are level with the tree top.

a) If he is 6' tall, how tall is the tree?

b) How does his answer change if the slope is actually 25° and not 30°?

c) 35° instead of 30°?

## Trigonometry

"Trigonometry" exposes students to real-world phenomena which require the use of a variety of trigonometric functions.

This lesson consists of thirty-eight applications of trigonometry in a wide variety of fields. Real-world applications in diverse areas such as photography, astronomy, geography, sailing, carpentry, sports and surveying give students a rich understanding of the uses of trigonometric and circular functions in problem solving. In the carpentry problem (shown here), one of the simpler exercises, the students are told that they have a summer job putting up outhouses in county parks and must precut the rafter ends so that they will be vertical when placed. The students must figure out the angle at which to cut the rafters.

In this lesson calculator use is optional.

- For further information about the *Sourcebook* turn to pages 58 and 113.

From Sourcebook of Applications of School Mathematics, "Trigonometry and Logarithms," pages 207-247. Prepared by a Joint Committee of the MAA and NCTM and available from NCTM.

*Statistics includes such activities as collecting, representing and processing data for simulations and/or samplings, fitting curves, testing hypotheses, and drawing inferences. The lessons below illustrate the varied and wide use of statistical analyses and inferences in our society.*

## Quantitative Literacy Series (QLS)

### Ice Cream Cone Prices

In "Ice Cream Cone Prices" students are asked to construct and draw inferences from a chart that summarizes data and to apply measures of central tendency, variability, and correlation.

Students analyze the data from a table of prices of single-scoop ice cream cones from 17 different stores. They are asked to create a stem-and-leaf plot of the prices. They look for gaps in the prices; they find the median and mean prices of an ice cream cone; and they look for the lower and upper quartiles and outliers of the data. Using these measures, they find and interpret the center, spread, and extreme values of the set of data, developing a richer understanding of statistical claims.

Calculator use is optional in this lesson.

- For further information about QLS turn to pages 48, 50, 79 and 109.

From Quantitative Literacy Series, Exploring the Data, Section III: Median, Mean, Quartiles, and Outliers, "Ice Cream Cone Prices," pages 48-49. QLS materials were written by members of the Joint Committee on the Curriculum in Statistics and Probability of the American Statistical Association and the National Council of Teachers of Mathematics, and is available from Dale Seymour Publications.

---

**Application 11**

**Ice Cream Cone Prices**

In September 1985, the prices of a single-scoop ice cream cone at 17 Los Angeles stores are given in the table below.

| Store (brand) | Price |
|---|---|
| Andi's (homemade) | $ .90 |
| Baskin-Robbins | .75 |
| Carvel | .95 |
| Cecelia's (Dreyers) | .90 |
| Cinema Sweet (homemade) | 1.20 |
| Clancy Muldoon | .95 |
| Creamery (homemade) | 1.05 |
| Farrell's | .70 |
| Foster's Freeze | .53 |
| Haagen-Dazs | 1.10 |
| Humphrey Yogart | .95 |
| Leatherby's (homemade) | .91 |
| Magic Sundae (Buds) | .96 |
| Robb's (homemade) | .95 |
| Swensons | 1.00 |
| Thrifty Drug | .25 |
| Will-Wright's (own recipe) | 1.15 |

1. Make a stem-and-leaf plot of the prices.

2. Are there any gaps in the prices? Where?

3. Find the median price of an ice cream cone using the stem-and-leaf plot.

4. Find the mean price of an ice cream cone.

5. Thrifty Drug's cone is much cheaper than the others. If it is taken off the list, do you think the median or the mean will increase the most?

6. Cross Thrifty Drug's price off the list before determining the following:

   a. Find the median price of the remaining cones.

   b. Find the mean price of the remaining cones.

   c. Which increased more, the median or the mean?

7. Find the range in prices. (Include Thrifty Drug from exercise 7 through 13).

8. Find the lower quartile of the prices.

# Introduction to College Mathematics (ICM)

## Leslie Matrix Model

The "Leslie Matrix Model" helps students examine a complex statistical problem by using given samplings and data. It further enables them to transform data and interpret and communicate statistical projections.

The future of Social Security, the future of veterans' benefits, and the changing school population in different regions are current issues in public policy that present age-specific growth questions. This model permits the examination of growth and decline of future populations within various age groupings. The model uses a matrix which includes information on birth and death rates for a population. Students are asked to use this matrix to determine long-term population levels. Several recurrence relations are used in developing this model.

- For further information about ICM turn to pages 64 and 97.

From Introduction to College Mathematics (ICM), Contemporary Precalculus Through Applications, Functions, Data Analysis and Matrices, "Leslie Matrix Model," pages 267-273. The program was developed by the Department of Mathematics and Computer Sciences, North Carolina School of Science and Mathematics, and is available from Janson Publications, Inc.

## 5   The Leslie Matrix Model

Population growth is a significant phenomenon for which many mathematical models have been developed. A frequently used model is the exponential function

$$P(t) = P_0 e^{kt},$$

in which $P(t)$ is a population growing without limits. Constrained growth of a population can be modeled by the logistic growth function. The logistic equation is

$$P(t) = \frac{QM}{Q + e^{-kt}},$$

in which $M$ is the maximum sustainable population and

$$Q = \frac{P_0}{M - P_0}$$

is the ratio of the initial population to the room for growth. Both of these models are macromodels, meaning that the models consider the population as a whole. In this section, a micromodel is developed that allows us to investigate questions about the different age groups within an entire population.

### 5.1   Modeling Age-Specific Population Growth

The future of social security, the future of veteran's benefits, and the changing school population in different regions of the country are current issues in public policy. The principal question arising in each of these discussions is how many people will be of a certain age after a period of time. The total population can be modeled with the equations above, but macromodels provide little help in answering age-specific growth questions. We would like to be able to examine the growth and decline of future populations according to various age groups. The model developed in this section will enable us to make these age-specific projections.

A fundamental assumption we will use in our model is that the proportion of males in the population is the same as the proportion of females, an assumption largely justified for most species. Consider a female population of small woodland mammals; Table 3 gives the populations for 3-month age groups. The total population in each age group is assumed to be twice the female population. The life span of this mammal is assumed to be 15–18 months, so none advance beyond the final column in Table 3. Our primary task with this data is to derive a mathematical model that will allow us to predict the number of animals in each age group after some number of years. To proceed, we first need to know something about the birth rate and death rate for each age group, rates that vary with age for most animal populations.

| Age (months) | 0–3 | 3–6 | 6–9 | 9–12 | 12–15 | 15–18 |
|---|---|---|---|---|---|---|
| Number of females | 14 | 8 | 12 | 4 | 0 | 0 |

Table 3: A Population of Small Woodland Mammals

*The ability to make informed observations and predictions about the likelihood of events is a powerful tool. These lessons help to develop a student's concept of experimental and theoretical probability. They encourage the growth of a student's intuitive understanding and the ability to analyze situations and design appropriate simulation procedures.*

# Contemporary Applied Mathematics (CAM)

## Variable Arrivals and Service Times– Simulation

"Variable Arrivals and Service Times– Simulation" provides a number of rich examples in which students can use experimental and theoretical probability. It allows them to explore the representation of problems involving uncertainty. It illustrates the use of random numbers in simulating situations.

This complex model provides students with an opportunity to explore variable arrivals, variable service times and multiple servers. The model can be used to approximate service in a fast food restaurant, as shown on this page, as well as in many other situations such as the Christmas rush at a small post office. Students interpret random numbers according to a frequency diagram and use those numbers to simulate the queues.

• For further information about CAM turn to pages 61, 65 and 92.

From Contemporary Applied Mathematics, Queues, Part IV, "Variable Arrivals and Service Times – Simulation," pages 16-34. CAM materials were developed by Wayne Copes, William Sacco, Clifford Sloyer and Robert Stark, and are available from Janson Publications, Inc.

## Part IV
## VARIABLE ARRIVALS AND SERVICE TIMES—SIMULATION

### A. Random Number Tables

In the last section, we examined the situation in which customers arrive and are processed at constant rates. While there are many applications for which that "model" is acceptable, there are times when it is not acceptable. For example, customers do not arrive at a fast food restaurant at a rate of exactly one each minute, nor does it take the same amount of time to serve each customer. A comparison of the types of arrival rates for the situation portrayed in the last section and in this section is shown in Figure 1.

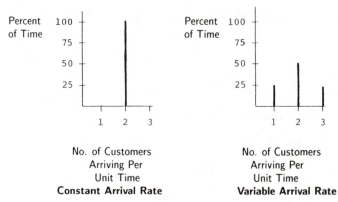

**Constant Arrival Rate**     **Variable Arrival Rate**

No. of Customers Arriving Per Unit Time

**Figure 1**

For the variable arrival rate situation, we do not know exactly how many customers will arrive during any unit of time. We only know that 25% of the time, 1 customer arrives in a unit of time; 50% of the time, 2 customers arrive per time unit; and 25% of the time, 3 customers arrive per time unit. It is usually assumed that the number arriving in one time unit does not depend on the number of arrivals in previous or subsequent time intervals. A similar situation exists in this section for processing or service times in that the required service time per customer can vary.

The *Standards* state that in grades 9–12 the mathematics curriculum should include the continued study of probability so that all students can–

- Use experimental or theoretical probability, as appropriate, to represent and solve problems involving uncertainty;

- Use simulations to estimate probabilities;

- Understand the concept of a random variable;

- Create and interpret discrete probability distributions;

- Describe, in general terms, the normal curve and use its properties to answer questions about sets of data that are assumed to be normally distributed;

**and so that, in addition, college-intending students can–**

- Apply the concept of a random variable to generate and interpret probability distributions including binomial, uniform, normal and chi square.

# Quantitative Literacy Series (QLS)

## Application 23

Step 1    Statement of the problem: Jon and Andy arrange to meet at the library between 1:00 and 1:30. They also agree to wait for the other person for five minutes. What is the probability that they will meet?

Step 2    Key component: The arrival times of Jon and Andy.

Step 3    Assumptions: Each of them arrives independently at some random time between 1:00 and 1:30.

Step 4    Model: Read two random numbers between 0 and 30. If the difference between these numbers is five or less, then Jon and Andy will meet; otherwise, they will miss each other.

Step 5    Trial: Read two random numbers between 0 and 30, and calculate the difference between them.

Step 6    Record the outcome of the trial.

Step 7    Repeat steps 5 and 6. You can also use SIMPRO12 to do this simulation.

See page 94 for a sample output using SIMPRO12 for this Application. The answers given here are based on this sample output. Answers will vary every time you run the program.

### Answers

1.  $\frac{17}{50} = 0.34$

## VI. SUPPLEMENTARY APPLICATIONS

Application 23

**Chances of Meeting**

Jon and Andy want to meet at the library. Each agrees to arrive there between 1:00 and 1:30 P.M. They also agree to wait five minutes after arriving (but not after 1:30). If the other does not arrive during that five minutes, the first person will leave. What is the probability that Jon and Andy will meet?

(*Hint:* Random times between 0 and 30 minutes can be selected from a random number table. Select two-digit numbers and eliminate those larger than 30.)

# Applications for Estimating Probability through Simulations

This lesson presents four situations in which students apply simulation techniques to estimate probabilities. Situations include:

- Chances of Meeting, pictured here, in which students estimate the probability of a meeting between two friends at a library within a half-hour period. They have agreed to wait for each other for five minutes.

- Making a Sale, in which students estimate the probability that a manager of a TV store will sell at least one TV during a certain hour.

- Back and Forth, in which students design a simulation and determine the average number of minutes it takes a child to reach either her home or playground.

- A Change in the Weather, in which students design a simulation to find weather patterns for July.

- For further information about QLS turn to pages 48, 50, 76 and 109.

From Quantitative Literacy Series, The Art and Techniques of Simulation, "Applications for Estimating Probability through Simulations," pages 45-48. QLS materials were developed by the Joint Committee on the Curriculum in Statistics and Probability of the American Statistical Association and the National Council of Teachers of Mathematics and are available from Dale Seymour Publications, Inc.

*Mathematics that handles properties of sets and systems with a countable number of elements, called discrete (discontinuous) mathematics, has emerged as a new aspect of the curriculum. These lessons illustrate some real-world applications of recurrence relations and other discrete structures.*

## High School Mathematics and Its Applications (HiMAP)

### Carbon Dating

In "Carbon Dating" students learn to represent a problem using a discrete structure such as a recurrence relation.

The lesson introduces students to a technique used by archaeologists to ascertain the age of ancient objects–carbon dating. The students are asked to determine the age of the paintings in the caves of Lascaux in France by means of a recurrence relation. To solve this problem they must know the ratio of radioactive carbon (C14), which decays in objects, to nonradioactive carbon (C12), which remains constant. They are told that the caves now contain 15% of the C14 that was present when the paintings were made. To estimate the age of the paintings they can use a recurrence relation. Logarithms and algebraic manipulations are also applied to various aspects of this problem.

- For further information about HiMAP turn to pages 71 and 96.

If $A$ is the amount of $C^{14}$ present at the time the paintings were drawn, then at the end of the first 5000-year period the amount of $C^{14}$ present, denoted by $A_1$, is $A_1 = .5444256A_0$. At the end of the second 5000-year interval the amount of $C^{14}$ present is given by $A_2 = .5444256A_1$. If $k$ represents the number of 5000-year intervals since the paintings were drawn, we have a recurrence relation for the amount present:

$$A_k = .5444256A_{k-1}.$$

To date the paintings, we need to find the value of $k$ for which $A_k = .15A_0$.

$$.15A_0 = A_k = .5444256A_{k-1}$$

$$= .5444256(.5444256)A_{k-2} = (.5444256)^2A_{k-2}$$

$$= (.5444256)^2(.5444256)A_{k-3} = (.5444256)^3A_{k-3} \quad . \quad .$$

The pattern continues, so that

$$.15A_0 = A_k = (.5444256)^jA_{k-j}$$

If $j = k$, $k-j = 0$, so $.15A_0 = A_k = (.5444256)^kA_0$. Thus, we are looking for that power $k$ of .5444256 such that $.15 = (.5444256)^k$. Using logarithms:

$$\log(.15) = k\log(.5444256) \quad \text{so that} \quad k = \frac{\log(.15)}{\log(.5444256)} \sim 3.12$$

Therefore the paintings in the caves are approximately 3.12(5000) = 15,600 years old.

*Exercise 1*

In a certain country the population increases each year by an amount equal to 2% of the population at the beginning of the year. Suppose the present population is 25 million. Find a recurrence relation for the yearly population figures for the country. What will be the population 10 years from now?

From HiMAP Module 2, <u>Recurrence Relations</u>, Section 3: "Counting Backward; Carbon Dating," pages 8-12. HiMAP materials were developed by, and are available from, COMAP, Inc.

# Applications in Mathematics (AIM)

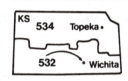

within the LATA. There are no routing decisions to be made here. The computer recognizes that the call is taken care of by the designated intraLATA carrier and it knows how to handle that.

This can be diagrammed in a very simple way:

```
                    LATA        TEL#
800-USE-MATH        532----------316-689-AAAA
```

## C.  The Decision Tree

Suppose the customer wants to receive calls at two different numbers (at a business phone during business hours and at a home phone during evenings and weekends). In this case the computer must make a decision based on the day and time of day. This is just the kind of thing a computer can do well provided it is told exactly how to make this decision in all possible cases.

EXAMPLE 2

The customer has intraLATA service in LATA 532. The customer wants its "800" calls to be received at 316-689-AAAA on weekdays from 9 to 5, and at 316-689-BBBB on evenings and weekends. The diagram indicating this situation looks like this:

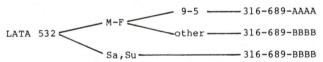

```
                           9-5 ------- 316-689-AAAA
                   M-F <
LATA 532 <              other ------- 316-689-BBBB
                   Sa,Su ------------- 316-689-BBBB
```

A branching diagram of this type showing the decisions that must be made is called a **decision tree**. Since the diagram shows the decisions needed in routing a telephone call, it may also be called a **call routing diagram** or a **routing tree**.

This particular tree is rather simple. Trees can become quite complicated as more and more decisions must be made. The decision tree is the best way to understand the logical sequence of decisions and to check the completeness and accuracy of a set of choices. Is every day of the week accounted for? Is every hour of each day included? Notice that for Saturday and Sunday no choice of hours needs to be listed.

An alternate form of the decision tree, useful for entering information into a customer record, is a **decision matrix**. A matrix is a rectangular array. Often it is an array of numbers but in this context it is an array of words which describes the decision process.

## Routing Telephone Service

In "Routing Telephone Service" students learn to represent and solve problems by using tree diagrams and matrices.

The materials produced under the AIM project are based on industry-related applied mathematics problems. They provide students with an experience in using their reading, writing and mathematical skills to solve real-world situations.

In this lesson, students must develop a routing plan and the appropriate decision tree for the installation of an "800" number, taking into account such factors as carrier choice, time choice, costs and service area information. They are asked to write a technical report on the problem and its solution. Additional problems and projects, detailed solutions, and a code for a program that computes the cost of various solutions are included.

Three videos that present the problem and its solution, a student resource book, a teacher resource book, and a microcomputer diskette are included in the learning module.

- For further information about AIM turn to page 90.

From AIM, "Routing Telephone Service," pages 1-57. AIM materials were developed by, and are available from, the Mathematical Association of America.

*The central ideas of calculus can be approached through informal activities that focus on the understanding of the concepts and their interrelationships. The lessons below are aimed at developing the conceptual understanding of such key ideas as infinite sequences and series and the corresponding associated limiting processes and the area under a curve.*

# Ohio State University Calculator and Computer Precalculus Project (C2PC)

## Maximum and Minimum Values

In "Maximum and Minimum Values" students determine the max-min points of a graph and interpret the results in problem situations.

In this lesson, local maxima and minima are introduced, and the general behavior of polynomials of degree three, four and five are discussed. Students use graphing calculators to find the local maxima and minima of a variety of polynomial functions. Applications to building and physics are included.

- For further information about Ohio State University Calculator and Computer Precalculus Project turn to pages 68, 84 and 107.

From Ohio State University Calculator and Computer Precalculus Project, Precalculus Mathematics: A Graphing Approach, Section 3.5, "Maximum and Minimum Values," pages 200-205. Materials were a collaboration between Ohio State University and three Ohio school districts, and are available from Addison-Wesley Publishing Company.

### 3.5   Maximum and Minimum Values

Many problem situations require that we find the largest or smallest value of a given function over some portion of its domain. We have encountered such situations in some of the real-world applications in previous sections. In this section we use zoom-in to find the coordinates of points where maximum and minimum values occur. Applications whose solutions involve finding a largest or smallest value of a model are also investigated.

- EXAMPLE 1: Draw a complete graph of $f(x) = x^3 - 4x$.

  SOLUTION: The graph of $f(x) = x^3 - 4x$ in the viewing rectangle $[-6, 6]$ by $[-10, 10]$ is shown in Figure 3.5.1 and is a complete graph.

  The graph of $f(x) = x^3 - 4x$ has a high point (peak) between $x = -2$ and $x = 0$. The value of the function at this point is called a local maximum value of $f$. Similarly, the graph of $f(x) = x^3 - 4x$ has a low point (valley) between $x = 0$ and $x = 2$. The value of $f$ at this point is called a local minimum value of $f$. More precisely we have the following definitions.

- DEFINITION   The value of $f$ at $x = a$ is called a **local maximum value** of $f$ if there is an interval $(c, d)$ with $c < a < d$ so that $f(x) \leq f(a)$ for all $x$ in $(c, d)$. The value of $f$ at $x = b$ is called a **local minimum value** of $f$ if there is an interval $(c, d)$ with $c < b < d$ so that $f(x) \geq f(b)$ for all $x$ in $(c, d)$.

  When we use the phrase "for all $x$ in $(c, d)$" in the above definition, it is implicitly understood that we mean those $x$'s in $(c, d)$ that are also in the domain of $f$. For example, using the above definition with this understanding about domain, the function $f(x) = \sqrt{x}$ has a local minimum value at $x = 0$.

- EXAMPLE 2: Determine the local maximum and local minimum values of the function $f(x) = x^3 - 4x$.

  SOLUTION: Refer to Figure 3.5.1. We use zoom-in to trap the low point between $x = 0$ and $x = 2$ in a very small viewing rectangle. The graph in Figure 3.5.1 suggests that we next look at the graph of $f$ in the viewing rectangle $[0, 2]$ by $[-4, -2]$ (Figure 3.5.2).

The *Standards* state that in grades 9–12 the mathematics curriculum should include the informal exploration of calculus concepts from both a graphical and a numerical perspective so that all students can–

• Determine maximum and minimum points of a graph and interpret the results in problem situations;

• Investigate limiting processes by examining infinite sequences and series and areas under curves;

**and so that, in addition, college-intending students can–**

• Understand the conceptual foundations of limit, the area under a curve, the rate of change, and the slope of a tangent line, and their applications in other disciplines;

• Analyze the graphs of polynomial, rational, radical, and transcendental functions.

# Project Calc

## Calculating Areas

In "Calculating Areas" students are asked to find the area under a curve, the rate of change, and the slope of a tangent line for a variety of curves.

This lesson is part of an introduction to integral calculus and asks students to determine the approximate area under a curve to a given degree of accuracy using rectangular slices. Students learn to improve the accuracy of these approximations by using larger and larger numbers of narrower and narrower slices. They also learn how to find the distance traveled by a moving object by finding the area under the velocity curve. And finally, they learn to use formulas for curves in conjunction with the rectangular slice approximation to make accurate calculations of areas and other quantities.

Several lab and computer problems, including using a balance to compute areas, are included.

• For further information about Project Calc turn to page 108.

From Development and Use of Modular Materials in Intuitive-Based and Problem-Based Calculus (Project Calc), Integration: "Calculating Area," Module 1, Units 1-4, pages 53-72. Project Calc was first published by Education Development Center. For further information contact The Learning Team.

---

**OBJECTIVE 1:** *To be able to approximate areas under a curve to a given degree of accuracy by using rectangular slices. And to be able to improve the accuracy of these approximations by using a larger and larger number of narrower and narrower slices.*

**Areas and the staircase approximation . . .**

You probably know what is meant by the area of a given shape, and how to calculate areas at least of simple shapes such as rectangles and triangles. The writeup "Measurement of Area" appearing in the box may be helpful background material. The question we shall discuss here is that of calculating areas of less simple shapes. The circle is a good example of a familiar shape whose area we often must know. A clue as to how we might go about this is provided by the "staircase" approximation of the previous unit. Figure 1a shows a quarter circle of radius 2 in, and Figure 1b shows how we can approximate this quarter circle by a staircase (compare these figures with Figures 1 and 2 of Unit I-3). What we want to point out here is that we can associate rectangular "slices" with the staircase; this is shown in Figure 1c. The idea behind what we do in this unit is based on the fact that the sum of the areas of these slices is an approximation (how good remains to be seen) to the area of the quarter circle.

This idea, which as we will see underlies all of integral calculus, leads to another closely associated idea. This is that an approximation based on rectangular slices, as illustrated in Figure 1c, can be improved by using narrower slices and more of them. We offer no proof of this fact, but we will examine two pieces of rather convincing evidence. The first of these is the set of diagrams shown in Figures 2a-f. Figure 2a is merely Figure 1c repeated, and shows an approximation to the quarter circle by 4 rectangular strips. In Figure 2b the number of rectangles

### CALCULATING AREAS

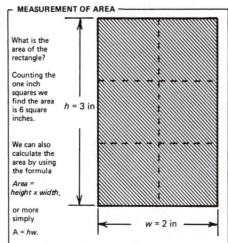

**MEASUREMENT OF AREA**

What is the area of the rectangle?

Counting the one inch squares we find the area is 6 square inches. $h = 3$ in

We can also calculate the area by using the formula

*Area = height x width,*

or more simply

$A = hw.$ $w = 2$ in

In this case we get $A = (3\ \text{in}) (2\ \text{in})$
$= (2 \times 3)\ (\text{in} \times \text{in})$
$= 6\ (\text{in} \times \text{in}).$

The units in the parentheses are better written as $A = 6\ \text{in}^2$ and read as "six inches squared." Sometimes, the unit one inch by one inch is called simply a "square inch" and is abbreviated sq. in. But the abbreviation $\text{in}^2$ is more compact, and indicates inches multiplied by inches, which is often useful. Use whichever term or abbreviation you like. We will usually write $\text{in}^2$, but speak of square inches.

When writing an area, or any other measurement, for that matter, it is essential to indicate the unit of measurement. (In this case, the unit is $\text{in}^2$.) Without the unit, the bare number tells us nothing. For example, if I tell you that my front lawn has an area of 100, I have not told you much. I could be speaking of 100 square feet (in the city); 100 square yards (in the suburbs); 100 acres (in the country); or even 100 square miles (in the Australian outback). To the statement, "The area is 100," the appropriate response is, "One hundred what?"

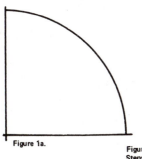

**Figure 1a.**

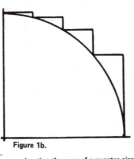

**Figure 1b.**

**Figure 1.**
**Steps in approximating the area of a quarter circle**

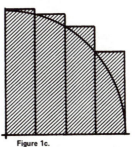

**Figure 1c.**

5E

*Mathematical structure provides the foundation on which the various content strands of mathematics are built. One way for students to begin to understand these underlying mathematical structures is by observing the common properties in systems that seem on the surface to be dissimilar. These lessons introduce students to the themes and logical consistencies found in the broad structuring principles of mathematics.*

# Ohio State University Calculator and Computer Precalculus Project (C2PC)

## Complex Numbers as Zeros

In this lesson students study the complex number system and its various subsystems with regard to its structural characteristics.

Complex numbers are introduced, and operations such as addition and multiplication are discussed. Students find complex number zeros in polynomials with real number coefficients. The fundamental theorem of algebra is stated and used to determine the number of zeros in a polynomial. Students develop a number of strategies for finding the zeros in a polynomial function. They also discover the limitations of graphing calculators for finding complex roots.

• For further information about Ohio State University Calculator and Computer Precalculus Project turn to pages 68, 82 and 107.

From Ohio State University Calculator & Computer Precalculus Project, Precalculus Mathematics: A Graphing Approach, Section 4.5,"Complex Numbers as Zeros," pages 273-284. Materials were a collaborative effort between Ohio State University and three Ohio school districts, and are available from Addison-Wesley Publishing Company.

## 4.5 Complex Numbers as Zeros

In this section we begin the study of complex numbers and find complex number zeros of polynomials with real number coefficients. The Fundamental Theorem of Algebra is stated and used to determine the number of zeros of a polynomial. Historically, complex numbers were introduced to provide solutions to equations such as $x^2 + 1 = 0$. This equation has *no* real number solutions. Mathematicians have constructed a number system that contains both the real numbers and the solutions to equations like $x^2 + 1 = 0$. This number system is called the complex numbers.

Complex numbers are important in real world applications as well as in pure mathematics. For example, complex numbers are involved in the study of problems in mechanical vibrations of objects, in the theory of alternating electrical current, and in the flow of fluids.

The synthetic division process explained in Section 4.3 is valid for complex numbers. However, the quotient and remainder can be polynomials with complex number coefficients in this case. The Remainder Theorem and the Factor Theorem are valid for complex numbers. These theorems and the division algorithm are valid for polynomials with complex number coefficients.

Let $i$ be a symbol with the property $i^2 + 1 = 0$. Formally, $i$ is a solution to the equation $x^2 + 1 = 0$. Thus, $i^2 + 1 = 0$ or $i^2 = -1$. We also write $i = \sqrt{-1}$. Both $i$ and $-i$ are solutions to the quadratic equation $x^2 + 1 = 0$.

---

• DEFINITION   An expression of the form $a + bi$, where $a$ and $b$ are real numbers is called a **complex number**. The set of all such expressions is called the set of **complex numbers**, or simply the **complex numbers**. The real number $a$ is called the **real part** of the complex number $a + bi$, and the real number $b$ the **imaginary part** of the complex number.

We agree that $a + 0i = a$. Then, the subset of complex numbers with $b = 0$ is just the set of real numbers. Before finding complex number zeros of polynomials we need to understand the arithmetic of complex numbers. Equality and the operations of addition, subtraction, multiplication, and division for complex numbers are defined as follows.

---

• DEFINITION   Let $a + bi$ and $c + di$ be any two complex numbers.

1. Equality: $a + bi = c + di$ if and only if $a = b$ and $c = d$.
2. Addition: $(a + bi) + (c + di) = (a + c) + (b + d)i$

The *Standards* state that in grades 9–12 the mathematics curriculum should include the study of mathematical structure so that all students can–

- Compare and contrast the real number system and its various subsystems with regard to their structural characteristics;

- Understand the logic of algebraic procedures;

- Appreciate that seemingly different

mathematical systems may be essentially the same;

**and so that, in addition, college-intending students can–**

- Develop the complex number system and demonstrate facility with its operation;

- Prove elementary theorems within various mathematical structures, such as groups and fields;

- Develop an understanding of the nature and purpose of axiomatic systems.

# Madison Project

## EXTENDING SYSTEMS: LATTICES. EXPONENTS

To begin with, we write numbers in an array or "lattice" like this:

```
                    ↑  (and so on)
31  32
21  22  23  24  25  26  27  28  29  30
11  12  13  14  15  16  17  18  19  20
 1   2   3   4   5   6   7   8   9  10
```

Now, this gives us a new way to write names for numbers:

(1) What number do you suppose is meant when we write

$$3 \rightarrow ?$$

(2) What number do you suppose is meant when we write

$$7 \uparrow ?$$

Can you find simpler names for each of these numbers?

(3)    8 →

(4)    9 ←

(5)    5 ↑ ↑ ↑

(6)    3 ↗

(7)    9 ← ↑ ←

[page 71]

(8)    21 ↓ →

(9)    3 ↑ ↑ ↑ ↑ ↑

(10)   3 ↑ ↑ ↑ ↑ ↓ ↓

(11)   24 ↙

(12)   26 ↙ →

(13)   27 ↙ → ↑

(14)   27 ↙ → ↑ ↑

(15)   27 ↙ → ↑ →

(16)   27 → → ↑ ↙

(17)   27 ↑ ↙ → →

First, we build up the "simple" or "basic" part of the structure. At this stage, Professor Page deliberately and wisely operates *on an intuitive level only*. He refuses to explain "how" he is doing these problems! To offer any explanation *at this stage* would make it nearly impossible for the children to use any creative originality in extending the system later on. Each child is invited to guess how *he* thinks these problems should be handled. Incidentally, it is probably a good idea to write the array

```
41  42  43                    . . .
31  32  33  34  35  36  37  38  39  40
21  22  23  24  25  26  27  28  29  30
11  12  13  14  15  16  17  18  19  20
 1   2   3   4   5   6   7   8   9  10
```

on the chalkboard at the front of the room, and keep it there during this entire discussion. But—*at this stage*—do *not* show how you are using this array!

(1)   **3 → names the same number that 4 does:**

$$3 \rightarrow = 4.$$

(2)   **7 ↑ names the same number that 17 names. Recalling the meaning of the symbol =, we can write**

$$7 \uparrow = 17.$$

(3)   **9**

(4)   **8**

(5)   **35**

(6)   **14**

(7)   **17**

(8)   **12**

(9)   **53**

(10)  **13**

(11)  **13**

(12)  **16**

(13)  **27**

(14)  **37**

(15)  **28**

(16)  **28**

(17)  **28**

Notice that in these problems we are *not* telling the students how to interpret the arrows—we are merely telling them the results of using the arrows. In particular, we are *not* telling them to interpret the arrows as "motions" on the array of numbers! (Actually, it is virtually certain that the students *are* interpreting the arrows in this way, but we are not allowing anyone to say so explicitly, because we shall soon want to ask the children just what the arrows really do mean.)

## Extending Systems: Lattices and Exponents

In "Extending Systems: Lattices and Exponents" students are given the opportunity to build two mathematical systems. This process helps to develop an appreciation of mathematical systems that seem different but share underlying structures.

In this lesson, students extend two mathematical systems: exponents and lattices. In both cases they are asked to develop the "basic" system first. When they run into the boundary of the system, they are asked to extend the structure consistently with what they have previously built. In this way they develop an intuitive feeling for the system and can further define the behavior, attributes and identities of the structures. For example, the system of integer exponents is extended to include zero, negative and fractional exponents.

- For further information about the Madison Project turn to pages 5, 14, 46, 73 and 101.

From the Madison Project, Explorations in Math, Lesson 24, "Extending Systems: Lattices & Exponents," pages 201-222. Explorations in Math was written by the Madison Project and is available from Cuisenaire Company of America.

**Section II - The Curricula**

# Overview of Curricula and Selection Process

The selection process for materials on the **MathFINDER CD-ROM** and this **Sourcebook** was complex, in good part because of the multiple goals of the Resources for Mathematics Reform Project. We wanted to provide an accessible archive of thousands of pages of materials which were supported in development by the National Science Foundation and are now out of print. In addition, we wanted to highlight a wide variety of materials – from the nineteen sixties through the present; from the U.S. and from other countries – that can support teachers' efforts to implement the *NCTM Standards*. And we wanted, if possible, to put all of our selected materials, along with the text of the *Standards*, on a single CD-ROM disc.

In the early weeks on the project, we searched through various sources to identify as many NSF funded mathematics materials as we could, going back to the earliest days of the NSF. We were aided in this collection effort by the generosity of numerous colleagues, who scoured bookshelves and basements to locate materials and make them available to us on loan. At the same time, we canvassed the members of our Oversight Committee to recommend other materials. We collected tens of thousands of pages in this way, and were forced to do some careful culling from the total collection. Members of the committee worked with us to judge materials according to such questions as:

- Are the materials currently available to teachers through some other means–for example, through a commercial publisher?

- How much do the materials seem to reflect the spirit of the *NCTM Standards?*

- When measured by the recommendations of the *Standards*, do the materials offer something that few other materials offer, for example, a special emphasis on mathematical modeling?

- How dependent are the materials on other materials (e.g., special manipulatives or software) that would not be included with the disc?

At the end of this long process, approximately 16,000 pages of materials were chosen and entered on the disc. Some of the older materials are included on the disc in their entirety. Other material – in particular, materials currently available through commercial publishers – are represented on the disc by a sampling of lessons.

We recognize that our selection process was not infallible and that we probably missed materials that some colleagues will wish we had included. In that case, we hope that others will follow in our footsteps and create new discs for the benefit of the thousands of teachers engaged in implementing the *Standards*.

*Mark Driscoll*
Education Development Center, Inc.

# Applications in Mathematics

*AIM offers students an introduction to problem situations taken from industry using an integrated curriculum package of video, computer software, and print materials.*

Each **Applications in Mathematics (AIM)** module features an on-site video for high school mathematics students in which an industry representative presents a problem situation and later leads the students to its solution.

Each of the seven modules may be used as a unit in a course, an independent study unit, a project for a math club, a tool in career counseling or a lecture in a workshop.

### A Backwater Curve for the Windsor Locks Canal
Kleinschmidt Associates is analyzing the Windsor Locks Canal to determine whether it can be used in the production of electricity. The president of the firm challenges students to find the upstream water surface elevation necessary to deliver 1,500 cubic feet of water per second to the downstream end of the canal.

### Budgeting Time and Money
AT&T analyst Arlene Taylor asks students to devise a plan to motivate workers to determine ways for work to be completed in a given length of time at the least cost.

### Capturing a Satellite
Astronaut George Nelson is to commute by Manned Maneuvering Unit (MMU) from the Space Shuttle to the Solar Max Satellite. The shuttle and the satellite are in the same circular parking orbit. Students must use orbital mechanics to determine how the MMU should be aimed in order to reach Solar Max with a minimum of in-flight adjustments.

### Pricing Auto Insurance
Insurance actuary Alfred Lichtenburg makes an analysis of receipts and expenses incurred in providing auto insurance. Students use data provided to calculate the fair price of auto insurance for specified age groups.

### Routing Telephone Service
Bell Communication researcher Dr. Susan Marchand asks students to devise a plan for routing a toll-free telephone number. They must consider time of day, costs, service area, etc.

### Testing Surface Antennas
Scientist Moray King is taking measurements of a new surface antenna. He asks students to figure a geometric adjustment so that the data can be interpreted relative to the center of the test antenna.

### Volcanic Eruption Fallout
Dr. William Rose, a volcanologist at Michigan Technological University, provides a problem based on distribution of fallout from eruptions at the Augustine Volcano. Students calculate predictions from a mathematical model and compare them to real data.

## Support
Each AIM module contains a:
- video cassette in which an industrial mathematician presents a real-life issue and solution
- student resource book detailing the problem
- teacher resource book describing how to use the unit
- computer diskette that explores the problem further through "What if..." questions

AIM videos, books and computer software are available on a free loan program. Users are encouraged to copy them and add them to their resource libraries.

## History
AIM, an MAA project conducted primarily by Oklahoma State University, was funded by a grant from the National Science Foundation and produced by Professors Jeanne Agnew and John Jobe.

© 1986 Mathematical Association of America.

## Available on MathFINDER CD-ROM
The print materials from all seven modules are contained on the MathFINDER CD-ROM.

## Order from
Mathematical Association of America
AIM Dissemination Clerk
1529 18th Street, NW
Washington, DC 20036-1385
Phone: 202/387-5200

## Examples
An example of an AIM lesson can be seen on page 81.

# Comprehensive School Mathematics Program

*CSMP is based on the belief that mathematics is a unified whole and should be learned as such. The program is organized in a spiral sequence so that students periodically return to concepts learned earlier.*

The purpose of the **Comprehensive School Mathematics Program (CSMP)** was to develop a K–12 mathematics curriculum to reflect the belief that the study of mathematics should be a challenging experience for all students. The result was the CSMP Elementary Program intended for students of all abilities in grades K–6.

The K–6 curriculum takes a spiral approach and includes a strong component of probability, statistics and combinatorics. A unique characteristic of CSMP is its use of non-verbal "languages" which give children immediate access to mathematical ideas and methods necessary for solving problems and expanding their understanding of mathematical concepts. The "languages" include the following: "The Language of Strings," based on set theory, used to organize and collect data; "The Language of Arrows," based on the notions of relations and functions, used to compare and analyze sets and their operations; and "The Language of the Papy Minicomputer," used to model the positional structure of our system of numeration.

The K–6 print materials include teacher guides, student worksheets, workbooks, and story books for the following programs:

- *CSMP Mathematics for Kindergarten*
- *CSMP Mathematics for First Grade: Parts I and II*
- *CSMP Mathematics for the Upper Primary Grade: Parts I-IV*
- *CSMP Mathematics for Intermediate Grades: Parts I-IV*
- *Activities for TOPS: A program in the Teaching of Problem Solving*

An earlier program developed by CSMP, *Elements of Mathematics Program* (7–12), explores the upper content limits of the mathematics that gifted students can understand and appreciate. The curriculum approaches both traditional high school mathematics topics and topics found in undergraduate mathematics courses from an advanced point of view. The content is treated in considerable depth and with rigor.

## Support
CSMP offers awareness sessions and 2-5 day training workshops.

Non-print materials include demonstration items, audio tapes, games, tools and manipulatives.

## History
CSMP was originally directed by Burt Kaufman and later by Clare Heidema. Funding came from the National Institute of Education and the US Department of Education.

© 1978-1986 CEMREL and McREL-Mid-continent Regional Educational Laboratory.

## Available on MathFINDER CD-ROM
A sampling of lessons from the CSMP Elementary Program are included on the MathFINDER CD-ROM.

## Order from
McREL
2550 South Parker Road, Suite 500
Aurora, CO 80014
Phone: 303/337-0990
Fax : 303/337-3005

## Examples
For examples of CSMP lessons turn to pages 11, 23, 26, 35 and 38.

# Contemporary Applied Mathematics

*CAM introduces students to mathematical topics with interesting and important real-life applications.*

**CAM 7-12+**

A stockbroker's portfolio, an exasperated customer at a crowded check-out counter, and a paramedic working under stressful conditions after a major earthquake are among the real-life scenarios presented in the **Contemporary Applied Mathematics** (CAM) series.

*Queues –Will This Wait Never End!*
It's Friday night at the local bank and the waiting line is trailing out the door... Shoppers are approaching a crowded check out area in a grocery store... These are two of several scenarios explored in this unit in which students learn to determine the balance between waiting time and costs. Simulations, a featured method of analysis, is shown to have many useful applications.

*Mathematics and Medicine*
A major earthquake has just struck San Francisco. Many residents of the city require medical attention. Where should the victims be taken? How can each patient's care be evaluated? Which patients should be treated first? How can one measure the severity of a patient's illness? This module teaches students to define, construct and apply mathematical indices that address these issues. Other mathematical concepts explored in this unit include relative information gain, curve fitting and logistic function, normalization and distance.

*Information Theory – Saving Bits*
By the year 2000, more than 40% of the work force will be involved in the collection, management and dissemination of information. This unit introduces students to information theory, the mathematical treatment of problems that arise in the storage and transmission of encoded information in binary form. It explains various types of error-detecting and correcting of codes.

*Graph Theory–Euler's Rich Legacy*
This unit introduces students to graph theory and its applications, and provides a history of the subject as well. Chromatic numbers, planarity, trees, and directed graphs are presented and utilized in numerous intriguing examples of problem solving.

*Glyphs–Getting the Picture*
In ancient Egypt picture writing, heiroglyphics was a key form of communication. Today scientists are discovering the value of modern picture writing, glyphs. This unit introduces students to this tool which can present multidimensional data in a meaningful and easy-to-interpret way. Also discussed are interesting applications of glyphs in medicine, astronomy, meteorology, geology, social studies and sports.

*Dynamic Programming–An Elegant Problem Solver*
Finding optimal solutions to problems, dynamic programming, is a powerful new method used by many people, such as investors looking for ways to enhance profits, physicians to treat patients, governments to reduce energy consumption, and coaches to achieve better training programs. The unit begins with a discussion of the shortest-path problem and proceeds to develop an appreciation of the efficiency of dynamic programming.

## Support
Each unit has a detailed teacher's manual which supplies:
- prerequisite and content information
- discussions to bolster the instructor's presentation of the material
- suggestions for additional activities related to each topic
- an appendix which defines terminology and discusses the mathematics

## History
CAM, a project of the Department of Mathematical Science at the University of Delaware, was written by Dr. William Sacco and Dr. Wayne Copes, industrial mathematicians at Tri-Delaware. CAM was funded by a grant from the National Science Foundation.

© 1988 Janson Publications, Inc.

## Available on MathFINDER CD-ROM
Seven sample lessons from *Graph Theory*, *Queues*, *Glyphs*, and *Mathematics and Medicine* are included on the MathFINDER CD-ROM.

## Order From
Janson Publications, Inc.
P.O. Box 6347
Providence, RI 02940-6347
Phone 1-800-322-MATH

## Examples
For examples of CAM lessons turn to pages 61, 65 and 78.

# Curriculum Development Associates Math

*The **CDA** mathematics curriculum reflects the developmental growth of children. Written language is kept to a minimum, making the program particularly suitable for multicultural situations.*

**Curriculum Development Associates Math (CDA)** provides three student mathematics booklets that promote multicultural learning. The books have as few written words as possible, so that students can focus on the math symbols and content. General directions are given in both English and Spanish.

Drills emphasizing quick recall of math operations are the backbone of CDA. Instruction on each major math concept is given in the following chronological order: expression of the idea through manipulatives, development of the idea with pictures, sketches and diagrams, and exploration of the idea in the abstract, using symbols.

CDA has two goals – to simultaneously increase the store of experiences that students can recall with ease and develop students' reasoning ability. With these goals in mind, sometimes problems are presented to students, and sometimes students pursue questions of their own design. As students work on problems, they try out solutions, refine guesses, formulate hypotheses, organize data and form generalizations. The open-ended nature of the curriculum enables teachers to adapt materials to learning objectives developed by school districts.

The major components of the program are:

* *The Think-Talk-Read (TTR) Center and Think-Talk-Connect Activities Kit*
  Materials which emphasize visual experiences and language development;

* *Patterns and Problems*
  Work books which encourage discussion of patterns and relationships;

* *Individualized Computation*
  Work books which emphasize computational fluency;

* *My Progress Book*
  Tests which correspond with the *Individualized Computation* materials;

* *Banking on Problem Solving*
  Activities which enable teachers to observe a number of responses, measuring students' problem-solving skills;

* *Drill and Practice*
  Supplemental materials which enrich and extend any basic program.

## Support

For parents, CDA supplies:
* *Making Friends with Numbers: Addition and Subtraction Kit*
* *Making Friends with Numbers: Multiplication and Division Kit*
* a letter to parents (available in English and Spanish) explaining the purpose of each section and suggesting activities that can be pursued at home.

For teachers, CDA supplies:
* *A Guide to CDA Math Curriculum*, which contains background papers explaining CDA's approach to teaching mathematics, readings for teachers, indexes to pupil and teacher materials, and placement activities.

In-service workshops are also available.

## History

The project was directed and materials were developed by Robert Wirtz from 1974-1982.

© 1974 Curriculum Development Associates, Inc.

## Available on MathFINDER CD-ROM

A sampling of lessons from CDA's *Drill and Practice* Activity pages are included on the MathFINDER CD-ROM.

## Order from

Curriculum Development Associates, Inc.
1211 Connecticut Avenue, NW, #414
Washington, DC 20036
Phone: 202/293-1760

## Examples

A CDA lesson can be seen on page 12.

# Developing Mathematical Processes

*With **DMP**, children explore their world, using manipulatives, then pictures, and finally abstract symbols to represent mathematical concepts.*

**DMP K-6**

The **Developing Mathematical Processes (DMP)** approach is based on the assumption that children learn best in an active environment, not one in which they are simply passive receivers of information, and that children develop a facility for abstract symbols by first learning about them in non-abstract situations.

The goals of DMP are:

- to foster more attention to the learning process and problems of the individual child;
- to develop more favorable teacher and student attitudes towards math;
- to help children become inquiring, independent problem solvers;
- and to enable children to acquire and demonstrate competency in the arithmetic of rational numbers, the geometry of physical time-space and the fundamentals of statistics.

DMP provides a complete mathematics program for grades K-6. Closely related specific learning objectives, and activities designed to promote their attainment, are clustered to form approximately 90 topics, each of which contains a teachers' guide, students' booklets, textbooks (for intermediate levels), testing materials, kits of activity materials, and printed materials. The topics are grouped in units. Non-print materials include games, lab equipment, filmstrips, films and slide tapes. Children learn by lectures, discussions, independent study, seminars, lab activities and demonstrations.

## Support

Each topic includes: multi-drawer kits containing story pictures, game boards, activity cards and demonstration graph paper, a teacher's guide, a student's booklet, pre and post tests.

In-service programs are available.

## History

DMP was begun in 1967 by Thomas Romberg and Harold Fletcher at the Wisconsin Research and Development Center for Cognitive Learning, University of Wisconsin-Madison. The program was designed by teachers, child psychologists and mathematics professors, and funded by the National Science Foundation, the US Office of Education and the National Institute of Education.

© 1974 The Board of Regents of the University of Wisconsin System for the Wisconsin Research and Development Center for Cognitive Learning.

## Available on MathFINDER CD-ROM

The eight topic booklets below are included:
Topic 15: Two Dimensional Shapes
Topic 48: Geometric Figures
Topic 56: Describing 3D Objects
Topic 63: Measuring
Topic 74: Decimal fractions
Topic: 86: Quadrilaterals and Other Figures
Topic 88: Ratios and Proportions
Topic 90: Patterns

## Order from

Delta Education
P.O. Box 915, Hudson, NH 03051
Phone: 800/258-1302
Fax : 603/595-8580

## Examples

For examples of DMP lessons see pages 19, 20, 40, 43 and 45.

# Discovering Geometry

*With **Discovering Geometry** students "create" geometry as they use an informal, guided discovery approach.*

**9-12**

**Discovering Geometry** is based on three ideas:

- students should be involved in exploring and testing geometric conjectures;
- only after students have sufficient grounding in the principles of geometry should they be introduced to the formal deductive aspects of the subject;
- students should be encouraged to work together.

In the *Discovering Geometry* textbook, students are not asked to memorize facts and formulas which can be remote and abstract but, rather, to explore geometric concepts and relationships using tools of geometry (compass, ruler, protractor and straightedge) and to construct their own definitions. For example, traditional texts tell students that measures of the angles of a triangle total 180 degrees. In *Discovering Geometry,* students are asked to measure the angles and determine the sum and to compare their findings with other students. Every lesson encourages interaction and communication. Students have an opportunity to write and talk about the mathematics they are encountering.

Many investigations in *Discovering Geometry* can be explored with the software tool *The Geometer's Sketchpad in* which students create and manipulate geometric figures and discover properties that hold even as a figure is distorted. *Exploring Geometry with The Geometer's Sketchpad* is a volume of over 100 reproducible computer activities correlated with *Discovering Geometry.*

## Support
- teacher's guide and answer key
- teacher's resource book
- several computer disks with Logo routines
- newsletter published twice yearly

## History
*Discovering Geometry,* published in 1989, is the product of Michael Serra's twenty years of classroom experience. *The Geometer's Sketchpad,* developed by Nicholas Jackiw of the Visual Geometry Project at Swathmore College, was funded by the National Science Foundation. *Exploring Geometry with The Geometer's Sketchpad* was funded in part by the National Science Foundation.

## Available on MathFINDER CD-ROM
Five sample lessons from *Discovering Geometry* are included on the MathFINDER CD-ROM.

## Order from
Key Curriculum Press
P.O. Box 2304
Berkeley, CA 94702
Phone: 800/338-7638
Fax:    510/548-0755

## Examples
For examples of *Discovering Geometry* see pages 62 and 70.

# High School Mathematics and Its Applications

*The **HiMAP** project provides a series of in-depth modules which both teachers and students find engaging.*

The eighteen **High School Mathematics and Its Applications (HiMAP)** modules are designed for both in-service and classroom use. Each module attempts to put mathematics in a real world context: an engineer employs algebra to construct a building, an astronaut uses error-correcting codes to communicate with mission control, an archaeologist investigates mathematical patterns to understand ancient cultures.

Each module contains between 40 and 90 pages and is supported when necessary by lesson plans, worksheets, transparencies, sample tests, computer software and suggestions for teaching strategies.

The modules are:

*The Mathematical Theory of Elections*

*Recurrence Relations –"Counting Backwards"*

*The Mathematics of Conflict*

*Symmetry, Rigid Motions, and Patterns*

*Using Percents*

*Problem Solving Using Graphs*

*Student Generations*

*The Appointment Problem: The Search for the Perfect Democracy*

*Fair Division: Getting Your Fair Share*

*A Mathematical Look at the Calendar*

*Applications of Geometrical Probability*

*Spheres and Satellites*

*The Mathematician's Coloring Book*

*Decision Making and Math Models*

*A Uniform Approach to Rate and Ratio Problems: The Introduction of the Universal Rate Formula*

*Exploresorts*

*The Abacus: Its History and Applications..*

*Codes Galore*

## Support

HiMAP modules may be supplemented by one or more of the following materials:

- videos
- computer software
- newsletters for teachers and pamphlets for advanced students
- lesson plans, worksheets, transparencies, sample tests
- calculator and computer applications.

## History

HiMAP modules were developed by COMAP, the Consortium for Mathematics and Its Applications. The HiMAP series was edited by Joseph Malkevitch of the City University of New York, and funding was provided by the National Science Foundation.

© 1984 COMAP, Inc.

## Available on MathFINDER CD-ROM

Sample lessons from six of the eighteen modules are included on the MathFINDER CD-ROM.

## Order from

COMAP, Inc.
Suite 210, 57 Bedford Street
Lexington, MA 02173
Phone: 800/77COMAP or 617/862-7878
Fax : 617/863-1202

## Examples

For examples of HiMAP lessons see pages 71 and 80.

# Introduction to College Mathematics

*Introduction to College Mathematics* challenges students to explore mathematics with the aid of calculators and computers.

ICM
11+

**Introduction to College Mathematics (ICM)** is a program designed to follow Algebra II for high school and college students. The goal of the project was to develop a syllabus and a few illustrative units for a new mathematics course that would be based on contemporary topics, making mathematics more relevant, and integrate calculators and computers into the teaching of mathematics.

Once the original syllabus and sample units were completed, they were developed into a fourteen-unit course focusing on mathematical modeling, computers and calculators as tools, applications of functions, data analysis, discrete phenomena and numerical algorithms. Three of the units, *Geometric Probability, Matrices,* and *Data Analysis,* are published by the National Council of Teachers of Mathematics (NCTM) in a series titled *New Topics for Secondary School Mathematics.* Introduction to College Mathematics materials have been published in textbook form by Janson Publications. The text is called *Contemporary Precalculus Through Applications: Functions, Data Analysis and Matrices.*

The materials are designed to stimulate students to become mathematical problem solvers, to communicate and reason mathematically, to value mathematics, and to develop confidence in their abilities. The lessons are application oriented and use technology to help students understand mathematical concepts. Interpretation skills are stressed throughout.

The course design requires access to technology, either graphing calculators or computers.

## Support
Support materials include:
• computer software
• answer key
• assessment package
• other teaching resources

## History
ICM was developed by the mathematics faculty at the North Carolina School of Science and Mathematics in 1985 with funding from the Carnegie Corporation of New York.

© 1988 NCTM and
© 1992 Janson Publications, Inc.

## Available on MathFINDER CD-ROM
Five sample lessons from the Janson *Contemporary Precalculus Through Applications* text, and one sample lesson from the NCTM *New Topics for Secondary Mathematics* series are included on the MathFINDER CD-ROM.

## Order from
Janson Publications, Inc.
P.O. Box 6347
Providence, RI 02940-6347
Phone: 800/322-MATH

NCTM
1906 Association Drive
Reston, VA 22091
Phone: 703/620-9840

## Examples
For examples from ICM lessons see pages 64 and 77.

# Journeys in Mathematics

*The **Journeys in Mathematics** materials enable teachers to guide students to generate their own knowledge of mathematics as they solve problems in engaging contexts.*

**Journeys K–8**

The goals of the **Journeys in Mathematics** project are to define and demonstrate a model for a new elementary mathematics curriculum that will prepare students for the challenges of the twenty-first century. The curriculum is designed to respond to the impact of technology on our society, take advantage of the educational potential of calculators, computers and video, reflect insights into children's learning from cognitive science and educational research, and help remedy the poor performance of American students on national and international assessments.

The activities found in *Journeys in Mathematics* reflect the underlying principle of the curriculum: the best way for students to learn mathematics is by *doing* mathematics. Each of the four *Journeys in Mathematics* modules encourages students to build mathematical connections. Problem solving, data collection, measurement and other mathematical concepts and skills are developed, practiced and applied.

The four modules are:

## What's Your Strategy?
This module helps students develop their problem-solving abilities, their confidence in approaching new problems, and their language for communicating and thinking about problem solving. It gives students experience in solving a variety of problems with coins, shapes and numbers.

## Computational Games
This module combines the playfulness of children's games with the intellectual challenge of mathematicians' games. Students will increase their computational proficiency, gain a better understanding of important concepts, and learn new skills.

## Data and Decisions
In this module, students make decisions and recommendations based on their analyses of sets of data. They learn to collect, organize, present and analyze information, and to integrate these skills into decision-making strategies.

## My Travels with Gulliver
This module integrates mathematics with reading, listening, writing and drawing. The module is based on the classic story *Gulliver's Travels*, by Jonathan Swift. The story and the unit describe Gulliver's voyages to Lilliput, the land of tiny people, and Brobdingnag, the land of giants.

## Support
Support materials include:
- teachers' guides
- student books
- manipulatives
- software
- letters to parents
- supplemental activities
- transparency and worksheet masters
- tests

## History
The *Journeys in Mathematics* project grew out of the Reckoning with Mathematics Project, funded by the National Science Foundation in 1986. The project was based at Education Development Center, Inc. (EDC), and was directed by Glenn Kleiman and Elizabeth Bjork.

## Available on MathFINDER CD-ROM
Sample lessons from each of the modules are included on the MathFINDER CD-ROM.

## Order from
Sunburst/WINGS for learning
1600 Green Hills Road
P.O. Box 660002
Scotts Valley, CA 95067-9908
Phone: 800/321-7511

## Examples
For examples of *Journeys in Mathematics* lessons see pages 4, 24 and 49.

# Lane County Mathematics Project

*The **Lane County Mathematics Project** materials emphasize the importance of problem-solving skills and encourage teachers to use a variety of instructional methods.*

**The Lane County Mathematics Project**, located in Lane County, Oregon, produced the *Problem Solving in Mathematics (PSM)* series, a program of problem-solving materials for grades 4–9. The project was developed in response to recommendations from the National Council for Supervisors of Mathematics (NCSM) and the National Council of Teachers of Mathematics (NCTM) that problem solving should be the focus of school mathematics for the 1980s.

Instructional strategies include direct instruction, guided discovery, laboratory work, small-group discussions, non-directive instruction and individual work. The packets are designed to be integrated into the regular mathematics program for each grade level.

The series consists of six individual books for grades 4–9, a book of alternative problem-solving activities for low achievers, and a guide and a set of audio tapes for in-service training. Each grade-level book contains approximately 80 lessons, a teacher's commentary with teaching suggestions and an answer key for each lesson. Each book begins with lessons that teach several problem-solving skills, proceeds with extensive drills and activities, and ends with a challenge section which encourages further development of these skills.

## Support
Support materials include:
- a textbook correlation chart done for the state of Oregon
- 100 pages of blackline masters per book
- teachers' guide
- *Alternative Problem Solving in Mathematics*–additional problems on which students work within specific time periods and guidelines
- nine in-service audio tapes
- in-service workshops

## History
The Lane County Mathematics Project was directed by Oscar Schaaf from the University of Oregon. The work was developed under an ESEA Title IVC grant from the Oregon Department of Education.

© 1983 Lane Education Service District

## Available on MathFINDER CD-ROM
A sampling of lessons from the grades four and five books and from the Algebra book are included on the MathFINDER CD-ROM.

## Order from
Dale Seymour Publications
P.O Box 10888
Palo Alto, CA 94303-0879
Phone: 800/USA-1100

## Examples
For examples of Lane County lessons see pages 6, 17 and 66.

# The Language of Functions and Graphs

*The Language of Functions and Graphs* aims to improve the performance of secondary-school students in interpreting and using information related to functions and graphs.

**The Language of Functions and Graphs** module, developed by the Shell Centre for Mathematical Education, grew out of the belief that while many students are well acquainted with graphs, tables of numbers, and algebraic expressions, and can manipulate them, they remain unable to interpret the global features of the information in them and are rarely given an opportunity to use mathematical representations to describe situations of interest to them.

The emphasis of the module therefore is on:

- Helping students to develop a fluency in using the mathematical language represented in graphs, tables and algebra to describe and analyze situations from the real world, and

- Creating a classroom environment that encourages thoughtful discussion as students try to comprehend or communicate information presented in mathematical form.

In the module, students are asked to translate information into various mathematical forms, translate between mathematical forms, and interpret them back into the situational context. To help develop these skills, students are encouraged to talk through their ideas and conceptions, present evidence and discuss explanations. It is even suggested that a discussion on "how to discuss" should introduce this module.

## Support
Support materials include:
- a video showing various teachers in the classroom, raising issues for discussion
- A microcomputer disk

## History
The Language of Functions and Graphs was directed by Malcolm Swan at the Shell Centre for Mathematical Education at the University of Nottingham in the United Kingdom. Funding was provided by the Shell Centre for Education and the Joint Matriculation Board.

© 1985 Shell Centre, Joint Matriculation Board

## Available on MathFINDER CD-ROM
Several sample lessons from *The Language of Functions and Graphs* are included on the MathFINDER CD-ROM.

## Order from
Shell Centre for Mathematical Education
University of Nottingham
University Park
Nottingham, UK NG72AD

## Examples
For examples of The Language of Functions and Graphs lessons see pages 37, 44, 60, 67 and 69.

# Madison Project

*The goal of the **Madison Project** was to create a mathematics curriculum that children would enjoy, that would produce a clearer conceptual understanding in students, and that would serve better than the traditional curriculum in preparing students for further study in later grades and in college.*

**Madison 2-10**

The **Madison Project** sought to combine clear notations, careful sequencing of mathematical topics, learning by generalizing from instances, and learning new concepts in a context providing appropriate tools for achieving a reasonable result.

The Madison Project produced print materials and films of classroom teachers using those materials. Lessons do not follow the standard progression, i.e., teachers telling students what will happen, showing them what to do, giving them practice or drills and, finally, summarizing the lesson. Instead, the Madison approach leads teachers to suggest one or more mathematical tasks and then work unobtrusively with students as they devise their own methods for tackling the tasks.

Project materials are supplementary and are best used in conjunction with other curricula to help build the classroom program.

## Discovery in Mathematics

This book encourages a three-ingredient approach to learning math – skills, ideas and uses – and proposes that all three ingredients be brought together to make a sound mathematics program. Its major area of concentration, however, is on the ideas of mathematics, observing and stressing the meaning of mathematical concepts, symbols and procedures.

## Explorations in Mathematics

This book is made up of supplemental materials which encourage a creative approach to the teaching of mathematics. In this approach an informal, non-authoritarian atmosphere is created in which a low anxiety level is achieved by the use of "non-lessons." The materials integrate exploratory lessons, discovery lessons, practice lessons, experience lessons, mastery lessons, and challenge lessons into a comprehensive approach.

## Support
Support materials include:
- a series of in-service films which show actual classroom lessons.

## History
Robert B. Davis began directing the Madison Project in 1957 with funding from the National Science Foundation and the US Office of Education.

## Available on MathFINDER CD-ROM
The MathFINDER CD-ROM contains six chapters from *Discovery in Mathematics* and the complete *Exploring in Mathematics* text.

## Order from
*Discovery in Mathematics* is available from Cuisenaire Company of America
12 Church Street, Box D
New Rochelle, New York 10802
Phone: 914/997-2600

## Examples
For examples of Madison Project lessons turn to pages 5, 14, 46, 73 and 85.

# Mathematics Curriculum and Teaching Program

*MCTP* is a collection of exemplary lessons from Australia that explore traditional mathematics topics in fresh ways.

**MCTP
K-8**

**The Mathematics Curriculum and Teaching Program (MCTP)** is a collaborative venture in Australian mathematics education. The project developers attribute the quality of the program to the collective wisdom of the teaching community in Australia, whose combined efforts to share their knowledge about teaching and learning formed the basis for the program. MCTP represents input from all of the Australian states and Territories, from both Government and non-Government sectors, and from teachers at all grade levels.

A major purpose of MCTP has been to find ways of sharing exemplary practice in the field of mathematics teaching. In particular, themes have emphasized the need for continued improvement in the effective use of technology, the potential benefits of cooperative group work and recognition that school mathematics has great relevance and purpose.

The MCTP style captures a spirit of excitement, which it attempts to inject into the classroom. Special attention is given to the way girls are portrayed in the math lessons.

One hundred fourteen lessons, divided into thirteen groups, are included in two MCTP *Activity Banks*. Each group contains an introductory section which discusses its educational value and five or six lessons that illustrate the theme of that group in the classroom. The thirteen groups are:

*MCTP Activity Bank, Volume 1*
• The role of video in the mathematics classroom
• Social issues in the mathematics classroom
• Concept learning – a first principles approach
• Physical involvement in mathematics learning
• Pupils writing about mathematics
• Mental arithmetic

*MCTP Activity Bank, Volume 2*
• Visual imagery
• Computers
• Estimation
• Story – shell frameworks
• Group investigation and problem solving
• Mathematical modeling
• Iteration – numerical methods

## Support
The classroom activities are supported by:
• worksheets
• cartoons
• computer software (IBM, Apple, Macintosh)
• practical tips
• a kit of assessment alternatives
• guidelines for consultants and curriculum leaders

## History
MCTP was conceived by the Curriculum Development Centre, Canberra, Australia

© 1988 Curriculum Development Centre

## Available on MathFINDER CD-ROM
Twenty-four of the 114 classroom activities are included on the MathFINDER CD-ROM.

## Order from
Curriculum Corporation
P.O. Box 177
Carlton South Victoria 3053 Australia

Materials can be ordered in the United States through:
NCTM
1906 Association Drive
Reston, VA 22091
Phone: 703/620-9840

## Examples
For examples of MCTP activities see pages 10, 16, 32, 51 and 74.

# Mathematics Resource Project

*MRP interweaves a series of support materials for teachers with ready-to-use classroom materials.*

**MRP
5-8+**

The **Mathematics Resource Project (MRP)** developed a set of teacher resource materials for middle school mathematics teachers. The materials were designed to help teachers improve their knowledge of mathematics, learn theories and teaching strategies, and help them locate and select materials to address the learning problems of their students.

The project developed five collections of in-service and instructional resource materials from which teachers could extend their knowledge and make the learning environment more flexible and effective. Each resource contains sections on mathematical content and didactics (information on learning theories and practices, techniques for diagnosing and evaluating, and alternative teaching strategies), along with comprehensive sets of classroom materials that develop skills and provide examples of applications and problem solving. Suggestions are given for student or class activities, starting points, carry-through and follow-up.

The five resource titles are:
* *Number Sense and Arithmetic Skills*
* *Ratio, Proportion and Scaling*
* *Geometry and Visualization*
* *Statistics and Information Organization*
* *Mathematics in Science and Society*

Each resource contains:
* Ideas on mathematical content
* Possible ways to extend and apply the topic
* Explanations, including information on learning theories and practices
* Techniques for diagnosing and evaluating
* Alternative teaching strategies
* Comprehensive sets of classroom materials that develop skills and provide examples of applications and problem solving, plus suggestions for student or class projects

## History
The Mathematics Resource Project was directed by Dr. Alan Hoffer at the University of Oregon in 1974, and was funded by the National Science Foundation.

©1978 University of Oregon

## Available on MathFINDER CD-ROM
Two sample lessons from *Geometry and Visualization* are included on the MathFINDER CD-ROM.

## Order from
Creative Publications
5040 West 111th Street
Oak Lawn, IL 60453
Phone: 800/624-0822

## Examples
For examples of MRP lessons turn to pages 34 and 52.

# Middle Grades Mathematics Project

*MGMP stresses the importance of clustering mathematical ideas and concepts to find the relationship among them.*

**MGMP 6-8**

The **Middle Grades Mathematics Project (MGMP)**, based at the Department of Mathematics of Michigan State University, represents a wide-ranging effort to improve the teaching and learning of mathematics in grades six, seven and eight. This effort included research on children's mathematics learning, the creation of "model" mathematics curriculum units, evaluation of those units and teacher in-service programs.

With MGMP, teachers launch a mathematical challenge to their students. Students work individually or in small groups to solve the problem, then return to a whole-class discussion to synthesize their experiences. The lessons are activity based and focus on problem-solving and critical thinking skills.

Each unit is made up of 8-10 activities and requires 2-3 weeks of instructional time. During the *launch* phase, the teacher introduces new concepts, clarifies definitions, reviews old concepts, and issues the challenge. During the *exploration* phase, small groups of students gather data, share ideas, look for patterns, make conjectures, or develop other problem-solving strategies. The teacher, by asking appropriate questions, encourages the students to persevere in seeking a solution to the challenge. During the *summarizing* phase, students return to a whole-class discussion to summarize the results of their data gathering. Through the summary, the teacher helps the students deepen their understanding of both the mathematical ideas in the challenge and the strategies used to solve it. A script is supplied to help the teacher in each phase of instruction.

The five units developed characterize this instructional model:

* *Mouse and Elephant: Measuring Growth*
  Provides experiences with area, perimeter, surface area and volume;

* *Factors and Multiples*
  Uses games, puzzles and problem situations to stimulate learning;

* *Similarity and Equivalent Fractions*
  Examines indirect measurement, scale models, and enlargement;

* *Spatial Visualization*
  Explores 3-dimensional objects in 2-dimensional drawings, then reverses the process;

* *Probability*
  Introduces concepts such as fair games, equally likely outcomes, expected value, simulations and binomial distribution.

## Support
Support materials include:
* blackline masters for worksheets
* teacher notes for each activity
* review problems
* unit tests and complete answers
* kits of manipulative materials for a class of 30 students: 600 wooden cubes in 6 colors, 800 tiles in 4 colors, 10 transparent 1/2- cm grids, 75 rubber bands, 8 plastic mirrors, 20 blank spinners, 15 pairs of dice, 18 ping pong balls, 1 roll of circular gummed labels, and 200 two-color counters

## History
MGMP began in the 1970s at the Department of Mathematics of Michigan State University. The series was funded primarily by a grant from the National Science Foundation.

## Available on MathFINDER CD-ROM
A sampling of lessons from the MGMP materials is included on the MathFINDER CD-ROM.

## Order from
Addison-Wesley Publishing Company
Attention Supplemental Publishing Div.
Route 128
Reading, MA 01867
Phone: 800/447-2226

Kits available from:
Cuisenaire Co. of America
P.O. Box 5026
White Plains, New York 10602-3142
phone: 800/237-3142

## Examples
For examples of MGMP lessons turn to pages 39 and 41.

# Minnesota Mathematics and Science Teaching Project

*MINNEMAST produced a coordinated curriculum in mathematics and science for grades K–3, as well as corresponding teacher preparation materials and teaching aids.*

**MINNEMAST K–3**

---

The specific objectives of the twenty-nine **Minnesota Mathematics and Science Teaching Project (MINNEMAST)** units were to develop process acquisition, attitudinal changes, organizational abilities and scientific literacy.

MINNEMAST writers believe that students should be shown as much of the larger picture as they can comprehend at their developmental level. They attempted to alleviate the fragmented student view of mathematics by providing models that would remain constant throughout the child's experience with the subject, even at the high school and college levels. Instruction methods include independent study, laboratory investigations, discussion, reading and writing.

The units are:

Unit 1:   Watching and Wondering
Unit 2:   Curves and Shapes
Unit 3:   Describing and Classifying
Unit 4:   Using Our Senses
Unit 5:   Introducing Measurement
Unit 6:   Numeration
Unit 7:   Introducing Symmetry
Unit 8:   Observing Properties
Unit 9:   Numbers and Counting
Unit 10: Describing Locations
Unit 11: Introducing Addition and Subtraction
Unit 12: Measurement with Reference Units
Unit 13: Interpretations of Addition and Subtraction
Unit 14: Exploring Symmetrical Patterns
Unit 15: Investigating Systems
Unit 16: Numbers and Measuring
Unit 17: Introducing Multiplication and Division
Unit 18: Scaling and Representation
Unit 19: Comparing Changes
Unit 20: Using Larger Numbers
Unit 21: Angles and Space
Unit 22: Parts and Pieces
Unit 23: Conditions Affecting Life
Unit 24: Change and Calculations
Unit 25: Multiplication and Motion
Unit 26: What Are Things Made Of?
Unit 27: Numbers and Their Properties
Unit 28: Mapping the Globe
Unit 29: Natural Systems

## Support

Support materials include:
- overview
- questions and answers about MINNEMAST
- student manuals for K-3
- teaching aids for K-1
- *Living Things in Field and Classroom,* a handbook for K-3
- *Adventures in Science and Math* – historical stories
- suggestions for programs for grades 4-6

## History

MINNEMAST was directed by James Werntz, Jr., Professor of Physics at the University of Minnesota School Mathematics and Science Center. The program was funded by the National Science Foundation from 1961-1970.

©1967 Regents of University of Minnesota

## Available on MathFINDER CD-ROM

All units except 15, 19, 26 and 29 are included on the MathFINDER CD-ROM.

## For further information, contact

The Learning Team
10 Long Pond Road
Armonk, NY 10504-0217
Phone:  914/273-2226
Fax :    914/273-2227

## Examples

For examples from MINNEMAST see pages 8 and 21.

# Nuffield Mathematics Teaching Project

*Nuffield* materials are aimed at teachers rather than students and stress how students learn.

**K-6**

The **Nuffield Mathematics Teaching Project** was directed, beginning in 1964, by Geoffrey Matthews at the Centre for Science Education in London. The materials stress how students learn, not what to teach. The central idea of the Nuffield activities is that children must be free to make their own discoveries and think for themselves, instead of learning from routine drills. This consideration of the child's cognitive development is reflected in the project's careful movement from pictorial to symbolic representation in all topics.

The materials developed by the Nuffield Project were aimed at teachers rather than students. The Nuffield approach is introduced in the book *I Do and I Understand*, and then explained in detail in the project's series of Teachers' Guides, Weaving Guides and Check-up Guides.

Teachers' Guides contain teaching suggestions, examples of apparently non-mathematical subjects and situations that can be used to develop a students' mathematical sense, examples of children's work, and suggestions for class discussions and out-of-school activities. The Weaving Guides are single-concept books that give detailed instructions or information about a particular subject. The Check-up Guides provide ways of measuring the children's progress. Films were also produced as part of the Nuffield Project.

## Support
- *I Do and I Understand*, a guide which focuses on the Nuffield approach
- Teachers' Guides focusing on: computation and structure, shape and size and graphs
- Weaving Guides presenting information and instructions on single concepts
- Check-up Guides offering teachers' evaluation methods that reduce the need for traditional tests. (Prepared at the Institut des Sciences de L'Education in Geneva under the general supervision of Piaget)

## History
The Nuffield Project, which was begun in 1964, was directed by Geoffrey Matthews at the Centre for Science Education in London, England. Funding was provided by the Nuffield Foundation.

© 1972 Nuffield Foundation

## Available on MathFINDER CD-ROM
Lessons from five of the Nuffield books are included on the MathFINDER CD-ROM.

## Order from
Nuffield-Chelsea Curriculum Trust
King's College
552 King's Road, London UK  SW10 0UA

## Examples
For examples from the Nuffield Mathematics Teaching Project see pages 13 and 15.

Translated into Italian, Dutch and French.

# Ohio State University Calculator and Computer Precalculus Project

*C2PC is a curriculum revision project designed to improve the mathematics preparation of college-bound high school students through the innovative use of technology.*

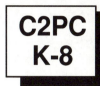

**The Ohio State University Calculator and Computer Precalculus Project (C2PC)** consists of a textbook and teacher resource guide, *Precalculus Mathematics: A Graphing Approach*, as well as computer graphing software. The textbook includes most of the traditional topics covered in a high school precalculus mathematics course. The most popular graphing software packages and/or graphing calculators can be used along with the book. It is unique in that the use of computer graphing capabilities or graphing calculators are assumed in the examples and exercises and must be used by students on a regular basis, both in class and for homework.

Computers and graphing calculators are used to quickly produce accurate graphs of functions, conic equations, polar equations, parametric equations, and surfaces (functions of two variables). Strong emphasis is placed on applications and on graphing both as a tool to build geometric intuition and to solve equations, inequalities, and max-min problems. This highly geometric approach to problem solving and precalculus mathematics is designed to strengthen students' understanding of mathematics and to increase the number of college freshmen prepared for collegiate mathematics study with precalculus or its equivalent.

## Support

Support materials include:

- The Graphing Calculator and Computer Graphing Laboratory student manual
- Instructor's Manual
- Master Grapher version 1.01 IBM, Apple II, and Macintosh

## History

The project was a collaborative effort between Ohio State University, three Ohio school districts, and many high schools throughout the United States that participated in field tests. Funding came from the Ohio Board of Regents, British Petroleum, and the National Science Foundation. The program was directed by Franklin Demana and Bert K. Waits from 1985-1989.

©1990 Addison-Wesley Publishing Company

## Available on MathFINDER CD-ROM

A sampling of nine lessons from *Precalculus Mathematics: A Graphing Approach* is included on the MathFINDER CD-ROM.

## Order from

Addison-Wesley Publishing Company
1 Jacob Way
Reading, MA 01867

## Examples

For examples from C2PC lessons see pages 68, 82, and 84.

# Project Calc
## Development and Use of Modular Materials in Intuitive-Based and Problem-Based Calculus

*Project Calc* makes calculus "student friendly" by incorporating instructional problems from magazines and texts.

**Proj Calc 10–12**

**Project Calc – Development and Use of Modular Materials in Intuitive-Based and Problem-Based Calculus** consists of a core course and a module bank. Graphical and numerical techniques are used to introduce concepts, making calculus accessible to a wide range of students. The materials facilitate the assimilation of these concepts by developing them from student activities, including hands-on laboratory and calculator/computer experiences.

Because modular materials address different levels, a variety of short courses for different audiences may be built from combinations of modules. This format also makes possible the use of individual modules as supplements to a standard course, or as substitute coverage for topics in a standard course. Typically, the presentation begins with a specific case and proceeds, in an interactive style, to generalizations and their applications.

Project materials include:

- *Module I: Integration*
- *Module II: Differentiation*
- *Module III: Relating Integration and Differentiation*
- *Resource Catalog*
- *Developer's Catalog*

## History
Project Calc was directed by William U. Walton at Education Development Center in 1973 and funded by the National Science Foundation.

## Available on MathFINDER CD-ROM
All three of the Project Calc modules are included on the MathFINDER CD-ROM in their entirety.

## For further information, contact
The Learning Team
10 Long Pond Road
Armonk, NY 10504-0217
Phone: 914/273-2226
Fax: 914/273-2227

## Examples
For an example from Project Calc see page 83.

# Quantitative Literacy Series

*QLS introduces statistics in the elementary and secondary curriculum.*

**QLS
K-9**

**The Quantitative Literacy Series (QLS)** consists of four books written by members of the Joint Committee on the Curriculum in Statistics and Probability of the American Statistical Association (ASA) and the National Council of Teachers of Mathematics (NCTM). The books are a result of a collaboration between statisticians and teachers who agreed on the statistical concepts most important for the general public to know and the best ways to teach these concepts. Five principles guide this series:

- There is often more than one way to approach problems in statistics and probability.

- Real data should be used whenever possible in statistics lessons.

- Topics traditionally taught in introductory statistics–such as standard deviation, normal distribution, hypothesis testing, and Bayes' Theorem and other probability formulas–should be taught after the more basic ideas in these four books.

- The emphasis in teaching statistics should be on good examples and on building intuition, not on showing how to lie with statistics or on probability paradoxes that may undermine a student's confidence.

- Students enjoy and profit from project work, experiments, and other activities designed to give them practical experience in statistics.

The four books are:

- *Exploring Probability*
- *Exploring Data*
- *Exploring Surveys and Information from Samples*
- *The Art and Technique of Simulation*

## Support
Support materials include:
- An in-service videotape providing an introduction to the project

## History
QLS was written by members of the Joint Committee on the Curriculum in Statistics and Probability of the American Statistical Association and the National Council of Teachers of Mathematics. Partial funding for the project was provided by the National Science Foundation.

© 1987 Dale Seymour Publications

## Available on MathFINDER CD-ROM
Nine sample lessons from the QLS books are included on the MathFINDER CD-ROM.

## Order from
Dale Seymour Publications, Inc.
P.O Box 10888
Palo Alto, CA 94303-0879
Phone: 1 800/USA-1100

## Examples
For examples from QLS lessons see pages 48, 50, 76 and 79.

# Regional Math Network

*RMN materials, written by classroom teachers, emphasize problem solving and problem posing.*

Each of the four **Regional Math Network (RMN)** books contains open-ended activities that encourage middle school students to pose their own problems for solving. Topics stress estimations, graphing, polling, interpreting charts, applications with calculators and computers and mental arithmetic. All of the materials include realistic mathematical applications that are accessible and motivating to students.

### Ice Cream Dreams (fractions)
*Ice Cream Dreams* is a collection of mathematical activities, games and manipulative materials developed to foster creative problem-solving skills. The activities involve work with fractions, decimals, percents, graphing, whole number computations, measurement and estimation.

### Math/Space Mission (estimation, geometry, relational concepts)
Students explore the outer realms of our solar system and learn estimation, geometry, relational concepts and problem solving. Module I presents introductory activities. In Module II students practice rounding numbers, estimating, graphing and problem solving as they explore the earth and its relationship to the sun and moon. Module III teaches ratio and proportion through activities about the space shuttle. Module IV provides an interactive experience, as students stretch their imaginations and use the skills they have learned to build a space colony.

### Quincy Market (ratio and proportion)
In *Quincy Market*, students use ratio and proportion, perimeter and area, linear measure, scale drawing and statistics to decide "Where should you set up shop?" and "What products should your business sell?" Students run their own business with either game cards or computer simulation. (Order disk separately.)

### Sport Shorts (decimals and percents)
*Sport Shorts* provides work on decimals and percents as students work with the records and statistics of their favorite sports teams. Students act as sports reporters and get their "stories" as they develop skills in problem solving, graphing, mental arithmetic, and analyzing data.

## Support
Each of the units contains a variety of teacher and student resources including teacher notes and teaching suggestions, student pages, answers, activity cards, transparency masters, manipulative material and classroom games.
In addition, the Quincy Market unit contains a computer disk suitable for an Apple Computer.

## History
The Regional Math Network project was directed by Katherine Merseth at the Harvard Graduate School of Education from 1985-1987. Funding was provided by the National Science Foundation.

© 1987 Dale Seymour Publications.

## Available on MathFINDER CD-ROM
Several sample lessons from *Math/Space Mission* and *Sports Shorts* are included on the MathFINDER CD-ROM.

## Order from
Dale Seymour Publications, Inc.
P.O. Box 10888
Palo Alto, CA 94303-0879
Phone: 800/USA-1100

## Examples
For examples from RMN lessons see pages 30 and 36.

# School Mathematics Study Group

*SMSG brings together educators, researchers, scientists and mathematicians for the development of math curriculum.*

The **School Mathematics Study Group (SMSG)** was a collaborative effort in the early 1960s involving college and university mathematicians, mathematics teachers, experts in education and representatives from the fields of science and technology. The primary purpose was to foster research and development in the teaching of mathematics by bringing together teachers and research mathematicians.

SMSG developed courses, teaching materials and teaching methods and produced materials for both students and teachers. Materials included elementary school texts for grades 1-6, junior high school texts, texts for slower students, senior high school texts for average and above-average students in college preparatory programs, supplementary and enrichment programs, technical reports, and case studies of SMSG teachers. Over 275 items were produced.

Many of the materials incorporate simple science experiments. The science texts are designed to be usable with a variety of mathematics textbooks. Previous acquaintance with science is unnecessary since the scientific principles involved are fairly simple and are explained, and scientific apparatus is kept to a minimum. Topics include linear functions, graphs, translation of axes, the distributive property, the solution of equations, and basic measurement of length, mass, time and temperature.

## History

SMSG was directed by Dr. E. G. Begle at the School of Education at Stanford University and was funded by the National Science Foundation. Parts of it became the New Mathematical Library, a publication series of the Mathematical Association of America.

## Available on MathFINDER CD-ROM

The MathFINDER CD-ROM contains, in their entirety, seventeen of the SMSG texts, including the *Mathematics Through Science* and *Mathematics and Living Things* series, three elementary school texts, and a sampling of senior high school texts that focus on geometry.

## Order from

The only SMSG material currently in print is *Mathematical Methods in Science,* which can be ordered from:

Mathematical Association of America
1529 18th Street, NW
Washington, DC 20036-1385
Phone: 202/387-5200

## Examples

Examples of SMSG lessons can be seen on pages 25 and 72.

# Secondary School Mathematics Curriculum Improvement Study

*The goal of the **SSMCIS** was to formulate, construct and test a unified secondary school mathematics program for the college-bound student.*

**SSMCIS
7-12**

**The Secondary School Mathematics Curriculum Improvement Study (SSMCIS)** was based on the assumption that the traditional four branches of secondary school mathematics– arithmetic, algebra, geometry, and analysis– could be taught as one unified mathematics. The goal of the project was to formulate, construct, and test a unified secondary school mathematics program for the college bound student in the upper 20% of academic ability.

The result was the development of a six-year mathematics program that unified instruction through fundamental concepts (set, relation, mapping, operation) and structures (group, ring, field, vector, space). The unified organization, along with the elimination of certain traditional topics, permitted the introduction of modern applications of probability, statistics, computer programming and linear algebra, as well as the usual application of differential and integral calculus in the 7-12 curriculum. The rigorous formality of the structures is not stressed- rather, their properties which shape the study of mathematics into a unified whole are emphasized.

The materials are available in six courses, one for each year in grades 7 through 12. Each course is accompanied by a teacher's commentary. Types of learning experiences include independent study, lectures, discussion sessions and computer-aided problem solving.

## Support
Printed materials include:
- objectives
- supplementary books
- teacher manuals
- tests
- newsletters.

## History
SSMCIS was conducted during the late 1960s and early 1970s by Howard Fehr at Teachers College, Columbia University, and funded by a grant from the National Science Foundation and the U.S. Office of Education.

© 1971 Teachers College Press

## Available on MathFINDER CD-ROM
The complete student text for all six courses is included on the MathFINDER CD-ROM.

## For further information, contact
The Learning Team
10 Long Pond Road
Armonk, N.Y. 10504-0217
Phone: 914/273-2226
Fax : 914/273-2227

## Examples
Examples of SSMCIS lessons can be found on pages 59 and 63.

# Sourcebook of Applications of School Mathematics

*The **Sourcebook** is a collection of applications in mathematics emphasizing the process of mathematical modeling which is recognized as the interface between mathematics and the real world.*

**7-12**

The **Sourcebook of Applications of School Mathematics** was compiled by a task force made up of mathemeticians, mathematics educators, representatives from industry and students from around the country. The objective was to develop and disseminate a comprehensive collection of examples of applications, suitable for grades 7-12, to address the lack of appropriate instructional materials. The committee found that most textbooks represented applications almost exclusively by artificial "story problems," which have a certain appeal as puzzles but are not likely to convince the skeptical student that mathematics has its uses in everyday life. The committee therefore selected lessons which emphasize the process of applying mathematics to real-world situations. As a result, students gain an appreciation of the importance of mathematics for our civilization, and are encouraged to apply mathematics in everyday life.

The core of the material is a collection of problems which are classified into three types:

- short problems similar in size and difficulty;

- medium-length problems that might serve as topics for a full class meeting;

- more time-consuming problems that could be used as bases for individual study projects.

Two introductory essays precede each problem: one from the viewpoint of an applied mathematician; the other from the viewpoint of a classroom teacher. The material includes a problem section (a collection of over 550 problems), a set of answers, an annotated bibliography, acknowledgments and a subject index.

## History

The Sourcebook of Applications of School Mathematics program was directed by Donald Bushaw in 1973 and prepared by a joint committee of the Mathematical Association of America (MAA) and the National Council of Teachers of Mathematics (NCTM). Funding was provided by the National Science Foundation.

© 1980 Mathematical Association of America (MAA).

© 1988 NCTM (New Topics)

## Available on MathFINDER CD-ROM

*The Sourcebook of Applications of School Mathematics* is contained in its entirety on the MathFINDER CD-ROM.

## Order from

NCTM
1906 Association Drive
Reston, Virginia 22091
Phone: 703/620-9840

## Examples

For examples from The Sourcebook of Applications of School Mathematics see pages 58 and 75.

# Teaching Integrated Math and Science

*With **TIMS**, a program that integrates math and science, children engage in activities resembling those of true scientists.*

The **Teaching Integrated Math and Science (TIMS)** project consists of a sequence of approximately eighty experiments that integrate math and science learning in the classroom. The experiments are adaptable to grades K-8 and are drawn from the physical and life sciences.

With TIMS, students work together to:

- identify variables, make observations and gather data in an organized way;

- make and check predictions and draw inferences and logical conclusions about the world in which they live.

TIMS is based upon a student-directed approach that depends on the guidance of informed teachers. The classroom becomes a laboratory setting in which children further their scientific understanding by investigating patterns in nature.

These experiments deal with a basic set of variables: length, area, volume, mass and time. Middle and upper-grade students eventually progress to more complex variables, including density, velocity, work and force. In doing the experiments, children in all grades learn to use the same method of investigation even in the simplest experiments: they draw a picture, make a data table, graph their results and answer questions addressing the concepts and principles demonstrated by the experiment. They learn to use the mathematics necessary to analyze and manipulate gathered data. Relying on readily available equipment and accompanied by detailed teacher preparation materials, TIMS experiments are practical for any classroom environment.

TIMS is most effective as a cohesive school-wide program that maintains thematic continuity from grade to grade as well as from one experiment to the next. TIMS advocates the team approach both for children and for teachers and their colleagues as they work with the project staff in implementing the program.

TIMS is currently being expanded so that a complete math and science curriculum will be available by 1995. Updated versions of many of the experiments are already available.

## Support
- Student pages for experiments
- Teacher materials
- TIMS Tutors
- Staff development workshops.

## History
TIMS grew out of Howard Goldberg's work with pre-service teachers beginning in 1976. The program is directed by Howard Goldberg and Phillip Wagreich at the University of Illinois at Chicago and funded by the National Science Foundation.

© 1985-1988 Howard Goldberg and Phillip Wagreich.

## Available on MathFINDER CD-ROM
A sampling of twelve TIMS experiments is included on the MathFINDER CD-ROM.

## Order from
TIMS Project, M/C 250
University of Illinois at Chicago
Box 4348
Chicago, IL 60680
Phone: 312/996-2448
Fax: 312/413-7411

## Examples
For examples from TIMS lessons see pages 2, 22, 47 and 54.

# Unified Science and Mathematics for Elementary Schools

*USMES* provides activities that integrate elementary mathematics, science and social science.

**USMES K-8**

**Unified Science and Mathematics for Elementary Schools (USMES)** provides 32 dynamic activities based on mathematical modeling. Each of the activities revolves around statements called "challenges" that present practical problems on which children can carry out in-depth investigations and acquire relevant skills. The challenges are based on the local school/community environment.

Development work, that is, setting up the challenge, is done by the classroom teacher. In response to the challenge, students decide on the course of action and are involved in all aspects of problem solving: observation, collection of data, representation and analysis of data, formulation and trial of successive hypotheses, and decision on final action to be taken.

With USMES, teachers and students develop the units together. By the time students and teachers have prepared the unit for implementation, the challenge has motivated students and encouraged good learning experiences.

During the challenge, students are engaged in independent study, programmed instruction, laboratory investigations, discussions, field experiences, problem analysis, quantification and experimentation (with the development of practical courses of action in mind), design, building, and investigation in the world outside the classroom. Directed learning materials help students solve technical problems.

Each group of students will respond to the same challenge differently. Student accomplishments determine the next phase of activity.

Over 30 units are offered, including: Bicycle Transportation, Soft Drink Design, Traffic Flow, Play Area Design and Use, and Consumer Research.

## Support

Each of the 32 units includes:

- edited log books that teachers wrote while they were conducting the challenge in their classrooms
- "How To" cards of student activities
- a "Design Lab" which discusses safety considerations and provides a list of supplies and hands-on activities
- a Curriculum Correlation guide which coordinates other curriculum materials with the USMES unit and helps teachers integrate USMES easily into other school activities and lessons
- newsletters
- in-service video series.

## History

USMES was begun in 1970 under the direction of Earle L. Lomon at Education Development Center, Inc. (EDC). Funding for the project was provided by the National Science Foundation.

© 1976 Education Development Center, Inc.

## Available on MathFINDER CD-ROM

A sampling of the "How To" cards and the teacher resource books for five of the USMES units is included on the MathFINDER CD-ROM.

## For further information, contact

The Learning Team
10 Long Pond Road
Armonk, NY 10504-0217
Phone: 914/273-2226
Fax: 914/273-2227

## Examples

For examples from USMES activities see pages 3, 31 and 55.

# University of Georgia Geometry and Measurement Project

*The goal of the **University of Georgia Geometry and Measurement Project** is to revise the elementary school geometry and measurement curriculum.*

**K-8**

To achieve its goal of revising the elementary school geometry and measurement curriculum for kindergarten through grade 6 the **University of Georgia Geometry and Measurement Project** developed 158 lessons, organized into the following three strands:

- Strand 1: Line and Length
- Strand 2: Angle; Surface and Area
- Strand 3: Space and Volume; Temperature; Weight

The lessons cover all traditional geometry and measurement topics except time and money.

Four basic ideas permeate the lessons: children can learn much more about geometry and measurement than they now learn in schools; children should learn geometry and measurement in the context of their personal activity; children should talk about the mathematical knowledge they are developing; teachers should be treated as professionals able to make their own curriculum decisions.

Each lesson includes detailed suggestions for activities and discussions, and many lessons include worksheets to help develop concepts and practice skills. Teachers are encouraged to adapt the lessons for use in their own classrooms.

Lessons are available either in print or on Macintosh diskettes. Each of the three diskettes includes a folder of lessons.

## Support

A network of former staff is available to conduct workshops on the use of the material.

## History

The University of Georgia Geometry and Measurement Project was funded in 1986 by the National Science Foundation and directed by James Wilson and William McKillip at the University of Georgia.

© 1990 University of Georgia

## Available on MathFINDER CD-ROM

All three Strands are included on the MathFINDER CD-ROM in their entirety.

## Order from

University of Georgia Geometry and
Measurement Project
Department of Mathematics Education
105 Aderhold Hall
University of Georgia
Athens, GA 30602
Phone: 404/542-4552
Fax: 404/542-4551

## Examples

For examples from the University of Georgia Geometry and Measurement Project see pages 9 and 18.

# University of Illinois Arithmetic Project

*The central theme of **UIAP** was that the study of mathematics should be an adventure for elementary school students, requiring and deserving hard work.*

**University of Illinois Arithmetic Project (UIAP)** developers asked what kind of mathematics do elementary students find interesting, and what has the most ultimate value for them? Their goal was not to develop a systematic curriculum but frameworks that provide day-to-day "here is something to try" ideas for the classroom. The project developers sought new ways to do old mathematics, that is, new structures or schemes within which could be found many interrelated problems revealing significant mathematical ideas.

The project was designed to convey both mathematics and pedagogy in an indirect way, to free teachers from the limitations inherent in any particular text or program, to enable teachers to capitalize on interesting ideas wherever they appear, and to encourage teachers to uncover and follow their own best instincts about what is interesting in math– in short, to teach the creative teaching of mathematics.

The project produced a book entitled *Number Lines, Functions, and Fundamental Topics*, three pamphlets: *Maneuvers on Lattices, Maneuvers on Number Lines*, and *Ways to Find How Many*; and twenty-five films and discussion materials as part of the K-6 mathematics framework.

## Support
Support materials include:
- 25 films and discussion materials for in-service and pre-service elementary teachers

## History
In 1958, UIAP was launched by David Page and Jack Churchill with Max Beberman at the University of Illinois. The program was funded by a grant from the Carnegie Corporation and moved to the Education Development Center, Inc. (EDC) in the mid 1970s.

©1969 University of Illinois Educational Services, Inc.

## Available on MathFINDER CD-ROM
The UIAP book and the three pamphlets – *Maneuvers on Lattices, Maneuvers on Number Lines* and *Ways to Find How Many* are included on the MathFINDER CD-ROM.

## For further information, contact
The Learning Team
10 Long Pond Road
Armonk, NY 10504-0217
Phone:  914/273-2226
Fax :    915/273-2227

## Examples
For examples from UIAP lessons see pages 27 and 42.

# University of Illinois Committee on School Mathematics

*UICSM encourages students to discover mathematics rules on their own and to use innovative problem-solving skills.*

**UICSM
6-12**

**The University of Illinois Committee on School Mathematics (UICSM)** is viewed as the progenitor of curriculum projects in mathematics in the 1960s, '70s, '80s and '90s. The UICSM was formed to investigate problems concerning the content and teaching of high school mathematics. The committee felt that a consistent high school curriculum had to answer the questions: What is a number? What is a variable? What is a function? What is an equation? and What is geometry? They also felt the need for precision of language and discovery of generalizations in the development of curriculum and methodology.

Until 1962, the UICSM program was devoted to producing a self-consistent and interrelated series of texts for college-bound students in grades 9-12. Beginning in 1962, the UICSM turned its efforts to the development of unusual approaches to topics in junior high school mathematics, appropriate for culturally disadvantaged students in large urban school systems.

The high school materials were divided into eleven units, covering algebra, plane Euclidean geometry, advanced algebra and circular functions, with tendencies on solid geometry, logic and other topics. A two-year vector geometry course was also produced. The UICSM 7th grade series, *Stretchers and Shrinkers*, developed the arithmetic of common fractions, decimals and percents, using unconventional models, with heavy reliance on pictures for students with low achievement in mathematics. The 8th-grade *Motion Geometry* series developed plane geometry through motions in the plane.

## Support
Support materials include a series of teacher-training films.

## History
Launched in 1951, UICSM was directed by Max Beberman at the University of Illinois. Funding was provided by the Carnegie Foundation, the U.S. Office of Education and the National Science Foundation.

## Available on MathFINDER CD-ROM
Included on the MathFINDER CD-ROM are the complete *Motion Geometry* series, and two of the four books in the *Stretchers and Shrinkers* series.

## For further information, contact
The Learning Team
10 Long Pond Road
Armonk, NY 10504-0217
Phone: 914/273-2226
Fax: 914/273-2227

## Examples
An example of a UICSM lessons can be seen on page 53.

# Used Numbers: Real Data in the Classroom

*The primary goal of the **Used Numbers** project is to help teachers and elementary school students become intelligent users of data.*

The **Used Numbers** project provides six units of study that teach elementary students how to collect and interpret data. These skills are traditionally taught at the secondary level but can be grasped by younger students as well.

In each unit, students collect data, record it in various ways, discuss and analyze it, learn that it can be interpreted in different ways, and make decisions based on interpretations of data. The lessons move gradually from short-focus activities to extended investigations. The primary-level books emphasize the building blocks of data analysis: counting, measuring and classifying. The upper-grade books continue to focus on real data and real problems while developing key concepts in statistics and graphing. Each book is approximately 100 pages.

## Counting: Ourselves and Our Families (K-1)
Students use counting to collect data, which they then sort, group, display and discuss. Activities include surveying student birthdays, counting students in class, guessing from clues the number of students in a class, counting noses and other body parts, and finding the ages of students.

## Sorting: Groups and Graphs (2-3)
Students collect a variety of materials, determine how to sort and classify them, and construct graphs and diagrams. Activities include classifying members of a "newly discovered life form," classifying neighborhood animals, and investigating scary things.

## Measuring: From Paces to Feet (3-4)
Students measure and investigate characteristics of themselves and the classroom environment. Activities include: "Robot paces," "Paces come in different sizes," "Are our feet a foot long?" "Using a smaller unit" (inches), "Classroom furniture: Do our chairs fit us?" and "How close can you get to a pigeon?"

## Statistics: The Shape of the Data (4-6)
An introduction to data analysis. Students begin with their own informal ways of describing data, then learn some of the formal measures such as medians. Activities include: "How many raisins in a box? " "How many people in our family?" and "How much taller is a fourth (fifth, sixth) grader than a first grader?"

## Statistics: Prediction and Sampling (5-6)
Students learn to make reasonable inferences from population samples. The activities introduce basic ideas about sample size, bias and making inferences. Activities include: "How a student sample describes a class," "Sampling family sizes," "Using color chips to make predictions," and "Cats: sampling a population".

## Statistics: Middles, Means, and In-Betweens (5-6)
Students learn how to interpret and use the mean, as well as other statistics, in the context of the data it represents. Activities include comparing and describing raisins, studying pulse rates, playing and studying a paper clip game, "The mean: another kind of middle," and "Means in the news".

## Support
Support materials include:
- a primary level in-service video
- a middle level in-service video
- ancillary items — plastic inch and foot sticks, Yektti Cards, cat poster

## History
Launched in 1986, the Used Numbers series is a project of the Technical Education Research Center (TERC), Lesley College, and the Consortium for Mathematics and Its Applications (COMAP). The materials were prepared with the support of the National Science Foundation.

## Available on MathFINDER CD-ROM
Sample lessons from three of the units are included on the MathFINDER CD-ROM.

## Order from
Dale Seymour Publications, Inc.
P.O. Box 10888
Palo Alto, CA 94303-0879
Phone: 800/USA-1100

## Examples
For examples from the Used Number turn to pages 7 and 33.

K-8

# ORDER YOUR MATHFINDER CD-ROMS AND ADDITIONAL SOURCEBOOKS HERE. PLUS, <u>FIND IT</u>!

## Find It! A Resource Guide to Hands-on-Math

Find It! is the most complete guide to available materials supporting hands-on teaching in mathematics. This guide is designed to spark ideas and save time in identifying the best manipulative-based materials for students' needs. It includes manipulatives, books, blackline masters, computer software, videos, films, toys, games, and picture books for grades pre-K-8 on all topics in math. Included is a list of sources and an annotated index of organizations and publications. 500 pages, 450 listings, over 100 sources. ISBN# 0-9624352-0-1

**The Learning Team**

Suite 350, 10 Long Pond Road, Armonk, NY 10504-0217

Please send _____ copies of the **MathFINDER CD-ROM** and **Sourcebook** package at $295 plus 5% shipping and handling. Please call for system requirements. Total $_____

Please send _____ copies of the **MathFINDER Sourcebook.** 1 - 9 $12.95 10 - 99 $10.95 ea. 100+ - $9.95 ea. Plus 8% shipping and handling with a $3.00 minimum. Total $_____

Please send _____ copies of **Find It!** 1 - $29.95 2 - 9 $26.95 ea. 10 + $23.95 ea. Plus 8% shipping and handling with a $3.00 minimum. Total $_____

Please enclose check or purchase order. New York State residents please include appropriate sales tax.

Name _____ Title _____

School _____

Address _____

City _____ State _____ Zip _____ Tel _____

For more information call us at 1-800-793-TEAM or FAX to 914-273-2227

---

**The Learning Team**

Suite 350, 10 Long Pond Road, Armonk, NY 10504-0217

Please send _____ copies of the **MathFINDER CD-ROM** and **Sourcebook** package at $295 plus 5% shipping and handling. Please call for system requirements. Total $_____

Please send _____ copies of the **MathFINDER Sourcebook.** 1 - 9 $12.95 10 - 99 $10.95 ea. 100+ - $9.95 ea. Plus 8% shipping and handling with a $3.00 minimum. Total $_____

Please send _____ copies of **Find It!** 1 - $29.95 2 - 9 $26.95 ea. 10 + $23.95 ea. Plus 8% shipping and handling with a $3.00 minimum. Total $_____

Please enclose check or purchase order. New York State residents please include appropriate sales tax.

Name _____ Title _____

School _____

Address _____

City _____ State _____ Zip _____ Tel _____

For more information call us at 1-800-793-TEAM or FAX to 914-273-2227

---

**The Learning Team**

Suite 350, 10 Long Pond Road, Armonk, NY 10504-0217

Please send _____ copies of the **MathFINDER CD-ROM** and **Sourcebook** package at $295 plus 5% shipping and handling. Please call for system requirements. Total $_____

Please send _____ copies of the **MathFINDER Sourcebook.** 1 - 9 $12.95 10 - 99 $10.95 ea. 100+ - $9.95 ea. Plus 8% shipping and handling with a $3.00 minimum. Total $_____

Please send _____ copies of **Find It!** 1 - $29.95 2 - 9 $26.95 ea. 10 + $23.95 ea. Plus 8% shipping and handling with a $3.00 minimum. Total $_____

Please enclose check or purchase order. New York State residents please include appropriate sales tax.

Name _____ Title _____

School _____

Address _____

City _____ State _____ Zip _____ Tel _____

For more information call us at 1-800-793-TEAM or FAX to 914-273-2227

If you want to stay on top of mathematics reform teaching for the nineties, in a manageable way, and meet the STANDARDS criteria for excellence — you need the MathFINDER CD-ROM.